GREAT PATENTS

ADVANCED STRATEGIES FOR INNOVATIVE GROWTH COMPANIES

GREAT PATENTS

ADVANCED STRATEGIES FOR INNOVATIVE GROWTH COMPANIES

David B. Orange

LOGOS
PRESS

GREAT PATENTS

ADVANCED STRATEGIES FOR INNOVATIVE GROWTH COMPANIES

David B. Orange

Published in The United States of America
by
Logos Press®, Washington, DC
www.Logos-Press.com
info@Logos-Press.com

10 9 8 7 6 5 4 3 2 1

ISBN-13
Hardcover: 978-1-934899-17-5
Softcover: 978-1-934899-18-2

Library of Congress Cataloging-in-Publication Data

Orange, David B.
Great patents : advanced strategies for innovative growth companies / David B. Orange.
p. cm.
Includes bibliographical references.
ISBN 978-1-934899-17-5 (hardcover) -- ISBN 978-1-934899-18-2 (softcover)
1. Patent laws and legislation--United States. I. Title.
KF3114.O73 2012
346.7304'86--dc23

2011029208

Contents

Author Contact Information

Chapter 1

David B. Orange

Patent Attorney

Buchanan Ingersoll & Rooney

1700 K St NW, Ste. 300

Washington, DC 20006

(202) 452-7977

david.orange@bipc.com

Chapter 2

Lawrence F. Rozsnyai

Senior Director, Intellectual Property and Legal Affairs

Metabolex, Inc.

3876 Bay Center Place

Hayward, CA 94545

(510) 293-8808

lrozsnyai@metabolex.com

Chapter 3

Oscar J. Llorin

Patent Attorney

The Law Office of Oliver Edwards LLC

8701 Georgia Avenue, Suite 605

Silver Spring, MD 20910

(301) 841-8050

ollorin@edwardslaw.pro

Chapter 4

John A. Evans

Senior Associate

Pillsbury Winthrop Shaw Pittman

2300 N Street, NW

Washington, DC 20037-1122

(202) 663-8096

john.evans@pillsburylaw.com

Chapter 5

Kevin W. McCabe

Director

Sterne Kessler Goldstein & Fox

1100 New York Avenue

Washington, DC 20005

(202) 772.8856

kmccabe@skgf.com

Chapter 6

William J. Bainbridge

Attorney

Cohen Mohr LLP

1055 Thomas Jefferson St NW

Washington, DC 20007

(202) 342-2550

bbainbridge@cohenmohr.com

CHAPTER 7

Marlin J. Blizinsky

Government Relations Officer
King County, Washington
2650 W. Newton St., Seattle, WA 98199
(206) 285-2653
marlinblizinsky@hotmail.com

CHAPTER 8

Luann Battersby

Patent Attorney, former scientist
Private practice primarily in pharmaceutical and estate planning issues
Box 215 20 W. Main St., Fairfield, Pennsylvania 17320
(301) 518-0423
Luann_Battersby@comcast.net

CHAPTER 9

Eric M. Fudo

BBP Partners, LLC
Washington, DC / Cleveland, OH
(202) 331-1154 / (216) 771-7943
efudo@bbppartners.com

CHAPTER 10

Leo J. Jennings

Counsel
Kramer & Amado, P.C.
1725 Duke Street
Alexandria, VA 22314
(202) 365-0443
leojennings@leojennings.com

CHAPTER 11

David O'Steen

President
Atlantic Intellectual Property Consulting, LLC
(202) 657-4482
dosteen@atlanticip.com

CHAPTER 12

Gerry Langeler

Managing Director
OVP Venture Partners
One SW Columbia Street, Suite 1675
Portland, Oregon 97258
(971) 327-2238
Langeler@ovp.com

Disclaimer

This book is designed to provide information about the subject matter covered. It is sold with the understanding that the publisher and the authors are not rendering specific legal advice within this book. The views of the authors may or may not reflect the views of their employers. This text is not a substitute for personalized advice from a knowledgeable professional. It does not create a lawyer-client relationship, and should not be relied upon. The publisher and authors cannot determine if the advice and strategies contained herein are suitable for your situation.

Every effort has been made to make this book as complete and as accurate as possible. However, there may be mistakes, both typographical and in content. Therefore, this text should be used only as a general guide and not as a replacement for advice from a professional you have engaged. While the publisher and the authors have used their best efforts in preparing this book, they make no representations or warranties with respect to the advice in this book. Neither the publisher nor the authors shall be liable for any loss or damage caused or alleged to be caused directly or indirectly.

Chapter 1

Introduction

David B. Orange, J.D.

Are patents not just some forms the attorneys fill out? As a patent attorney, I can tell when a patent has been drafted to sue industry giants, when it is intended to be a beacon for investors and other industry players, and when it just lacks an underlying strategy.

One of the most high-profile ways a patent can help a business is a big win in court. Patents, however, can also provide value by persuading your would-be competitors to pursue other opportunities, or at least by making them spend more on an opinion letter on your patent than you spent to get the patent. Working with high technology patents every day, I have seen this happen on many occasions, even if the patent owners were unaware of their victory. Patents can also bring in money through licensing, patent pools, and being a barrier to entry that you can show to investors.

In planning this book, the publisher and I agreed that there is a gap among books on patents between basic books which explain what a patent is, and advanced books written for practicing patent attorneys. Many of the successful executives I have met already deal with patents in one form or another, so they do not need a primer, but books on the theory of patent law would not be useful to them. This book aims to inform executives on how best to use patents to improve their competitiveness.

For all of the advantages patents can provide a company, one of the biggest challenges they present is that they are often planned and executed before they are used. The cash-flow sensitive leader is caught between having to spend money today to protect intellectual property, and having the protection of a patent when the technology changes the marketplace. This book tackles that challenge by focusing on insider advice, common pitfalls, expert strategies, and other wisdom that you can use today.

The chapters in this book are broken into two categories: Strategies and Valuation. While there is certainly overlap between these two topics, the goal is to acknowledge that companies pursue patents to make money, and at the same time to help companies discover what they can do with their patents that they may not have previously considered.

Central to both strategies and valuation is the idea that not all patents are created equal. Great patents reflect a business strategy. For example, breakthrough technologies that need to get to market quickly are often well served by the accelerated examination route which guarantees a final disposition in 12 months. However, for innovations that are more subtle, the more traditional approach, and its additional opportunities for the inventors to present their case, may be more appropriate.

My goal in editing this book was to wring as many actionable insights and quantified data out of each chapter author as I could. Reading this book cover-to-cover will put you out in front of your competition, but it is also designed so it can picked up for a particular chapter.

Any of the chapters could justify having this book on your shelf: If you are pitching to a venture capitalist, there is a chapter on working with venture capitalists (and it is written by a very successful managing director); if you happen to be in an industry with Standard Setting Organizations, you can read how one of Microsoft's top negotiators recommends maneuvering in Standard Setting Organizations towards your technologies and away from your competitors.

If you find a chapter that is relevant to your context—perhaps on advice for startups without in-house intellectual property counsel—

take a peek. You will find industry veterans with years of successful experience laying out the insights and know-how that set them apart.

Here is to your company's success!

David B. Orange
Washington, D.C.
2011

Section I

Strategies

Chapter 2

How a Young Company can Build and Manage its Intellectual Property

Lawrence F. Rozsnyai, Ph.D., J.D.

Lawrence Rozsnyai is currently the Senior Director of Intellectual Property and Legal Affiars at a clinical-stage pharmaceutical startup company in the San Francisco Bay Area, where he heads all intellectual property matters for the company. Dr. Rozsnyai began his legal career at an IP-focused national law firm where he wrote and prosecuted patent applications in the life sciences, medical device, and other technology areas. Dr. Rozsnyai's prior work experience includes business development at several biotechnology startups and management consulting with Fortune 500 clients. Dr. Rozsnyai received his J.D. from the University of Washington School of Law and a Ph.D. in chemistry from the Massachusetts Institute of Technology. Dr. Rozsnyai can be contacted at lrozsnyai@gmail.com.

This chapter discusses several concepts and recommends best practices that a company should consider adopting as it develops its intellectual property. While its primary audience is the CEO and senior management at a startup company progressing from its first few patent applications to building and managing a more robust portfolio, management at larger companies should also make sure their companies are following most of these guidelines.

This discussion assumes some working knowledge of the basic concepts of U.S. patent law. Concepts such as enablement, best mode, prior art and freedom to operate, should be familiar to you. Several accessible references explain these concepts very well. For a basic and practical guide, see D. Pressman, *Patent It Yourself: Your Step-By-Step Guide to Filing at the U.S. Patent Office*, Nolo (14th ed. 2009). For a more in depth survey of patent law, see J.M. Mueller, *Patent Law*, Wolters Kluwer Law & Business (3rd ed. 2009). And while the America

Invents Act, which is currently before Congress, may change some of these rules, most of the following basic concepts still apply.

YOUR FIRST FEW APPLICATIONS

Congratulations! You have your first invention and are ready to file your first patent application. You now have a few important decisions to make:

- How do we select our first IP counsel?
- Do we file a provisional application, or a "real" application? What is the difference?
- How do we go about drafting the first few applications?
- How much "stuff" should we include in our written description?

We now explore these questions and give some practical recommendations.

A. SELECTING IP COUNSEL

It is a good idea to retain a patent attorney fairly early on. Speaking with one or more attorneys before you file your first application will allow you to discuss your invention, related ideas, where you are going with your business, and will increase your understanding of how patents can support your business. In this way, you and the attorney can develop your initial strategy, decide what to put into the first application and how many to file, and consider longer-term issues. While it may cost you a bit of money up-front, asking intelligent questions and making some reasonable assumptions can save you much more over the next few years.

LAW FIRMS: BIG VS. BOUTIQUE

Large firms offer many resources to inventors. Many of the big firms have Ph.D. scientists, experienced engineers, or other subject matter experts who can write very good patent applications. Big firms can also help you with licensing and litigation concerns.

However, large firms may not be a good fit for individual inventors or small companies. Big-law partners can impress you with their Fortune 500 clients and experience, their ability to help you license your invention to other companies, and can tell you stories about their counterparts winning multi-million dollar patent suits. They will, of course, bill you for it: partners at large firms currently bill at about $550 to $650 or more per hour.

Therefore, know what you are getting into and why you are speaking with a big law firm. Do you need a litigation team now? Need help with licensing? Chances are you can cross these bridges at a later time.

Patent boutiques—smaller firms often specializing only in prosecution—can offer a good alternative, and are growing in popularity. Most often boutiques are started by a few big-law partners seeking a change in lifestyle, and perhaps a few in-house attorneys from large companies. They often specialize in specific technology areas such as internet technology and software, medical devices, biotechnology, mechanical devices, and so forth. While it can be more difficult to find a boutique in your technology than a large firm, many very good patent attorneys work at boutiques. Benefits include more "personalized" attention and lower rates: partners at boutiques tend to charge in the range of about $350 to $500, depending on location. Do not worry too much if your boutique does not do litigation because when the time comes, you can go talk to big law.

No matter which type of firm you select, remember that you are hiring a person more than a firm. Select the right individuals, not the name of the firm. Ask to meet the associates, since they are the ones who will be working all night on your cases. Ask for references. Make sure they know your technology area, or at least will not bill you too much to come up to speed.

Attorneys and Agents

According to the USPTO, "Inventors may prepare their own applications and file them in the Patent and Trademark Office and conduct the proceedings themselves, but unless they are familiar with these matters or study them in detail, they may get into considerable difficulty."[1] A non-inventor who files and prosecutes the application

1. http://www.uspto.gov/web/offices/pac/doc/general/attorney.htm (last visited on July 11, 2010).

on behalf of the inventor must be registered to practice before the USPTO. If that person is already an attorney, then she is called a patent attorney; if not, she is called a patent agent.

Although the scope of the work an agent can do is more limited than that of an attorney, working with an agent who is an experienced writer and arguer before the patent office—and who understands your technology—can be a very cost-effective option. Most patent agents should charge between the low $100s to the high $300s per hour depending on experience, and whether they are at a big firm, a boutique, or solo. If you really need to control your costs, consider working with a good patent agent. Then, if and when you need an attorney, you can hire one at that time.

B. PROVISIONAL APPLICATIONS AND "REAL" APPLICATIONS

We now turn to some of the "nuts and bolts" of patent law and writing patent applications.

WHAT IS A PROVISIONAL PATENT APPLICATION?

Since June 8, 1995, inventors have been allowed to file provisional patent applications with the U.S. Patent and Trademark Office (USPTO). Provisional patent applications were created to put U.S. inventors on equal footing with foreign inventors, who can first file a patent application under the Patent Cooperation Treaty, and secure that filing date when they file the same application in the U.S. up to 12 months later. A provisional application therefore gives U.S. inventors this same 12 month "head start" with the USPTO.

A provisional application is not a "real" patent application and differs from a real or "non-provisional" application in a few important ways:

- A provisional application is a separate application and will not turn into a non-provisional application. A provisional application will therefore never become an issued patent.
- A provisional application needs only to describe what your invention is. It needs no claims (more on this below).
- A provisional application expires in 1 year.

Why draft a provisional rather than a utility application?

A provisional application allows you to submit a sort of "rough draft" of what will eventually become the real (non-provisional) application, and gives you an earlier priority date. This earlier priority date basically gives you patent protection on the final claims that issue for 21 years rather than the 20 years you would get if you only file a non-provisional application.

Filing a provisional application is inexpensive. The current fee for an application owned by a "small entity" (a company with up to 500 employees[2]) is $110. This fee is doubled for non-small-entity companies.

> *File a provisional application. It is inexpensive and gives you an extra year of protection. A separate non-provisional patent application claiming priority to the provisional must be filed within 12 months, giving you some time to perfect your invention before the non-provisional application is submitted.*

C. Tips on Writing Your First Few Applications

As noted above, a provisional patent application needs to describe only what the invention is, what it does, and—if not generally known—how to make it. An inventor could literally submit a photocopy of her notebook with some supplemental explanation, and that would suffice.

However, hasty, sloppy—or merely inexperienced—drafting of a provisional application can wreak havoc on any non-provisional application that is eventually filed. It is much better to spend the time to write a proper application and avoid these traps from the start. Here are some recommend best practices to adopt during the drafting and writing stages that will increase the value of your first few patent applications.

Your Provisional Application Should Look Like a Real Application and Should Include Claims

While nearly anyone can write a competent provisional patent application, a non-provisional application based on the provisional application must be filed within 12 months. To preserve the priority date

2. To qualify for small entity status, your application must also not be licensed to a company larger than 500 employees.

of the provisional application, the corresponding non-provisional application cannot incorporate "new matter" that was not included in the provisional (see INSET: NEW MATTER). This means that even though your provisional application is a "rough draft" of the non-provisional application, you need to include *all* of the aspects of the invention in the first application (the provisional) for those aspects to receive that filing date. So in practice, you should write your provisional application so that your non-provisional application will basically be the same as your provisional application, at least as far as the written description is concerned. When you file the non-provisional application 12 months later, you can add claims and make minor changes in the written description, but most—if not all—of the provisional application will be repeated. If the provisional application is written well, writing the non-provisional application will be basically a "copy and paste" exercise.

Though not required, writing claims in the provisional application will force the inventor and drafter to make sure that the written description is both broad and specific enough to fully enable those claims. This is where photocopying your notebook will probably not suffice, since descriptions in notebooks are seldom broad or general enough to support claims that will cover parties beyond the most literal infringers of the narrowest aspects of the invention.

INSET: NEW MATTER

Once a patent application is filled, the text describing the invention (called the "written description" or the "specification") is, for the most part, fixed and cannot be changed. The reason has to do with the priority date of your application. When you file your application, the USPTO sets your filing date as the date of your invention (notebooks proving the date of inception can be used to get an earlier date, since the U.S. ultimately gives priority to this date).

Think of this priority date as a time stamp. Any new data or aspects of your invention can only be added as "new matter" and will, correspondingly, get a later priority date.

A "continuation-in-part" (CIP) application is an application that includes new matter, and therefore will have two priority dates—one for the original description, and one for the later-added text. It is often better not to file a CIP, and file a new patent application instead. Think of your first application as your "central" or "core" application, and ones containing new matter as new or improved aspects or applications of the core invention.

Avoid Common Pitfalls in Drafting

Writing a good patent application takes experience. A "good" patent application will not only meet the statutory requirements, but will avoid common pitfalls during prosecution. It will also minimize some common traps that could occur in litigation, should the patent that eventually issues ever be subject to litigation, either offensively or defensively.

Statutory Requirements	**Common Pitfalls**	
	During Prosecution	In Litigation
• Novel • Non-obviousness • Fully enabled • Discloses the best mode	• Providing insufficient support for fallback positions when broad claims are rejected • Vaguely describing non-core ideas	• Claims narrowed because of weak definitions or examples • Claims narrowed by "patent profanity" • Claims not practiced by one infringing party

1. Provide Sufficient Support for Fallback Positions

Your goal in patent prosecution is to get the broadest claims issued. This rarely happens. It is far more common to have your broadest claims rejected (usually in light of what is taught or considered obvious in the prior art) (see Inset: Prior Art), requiring you to retreat to some "fallback" position. If you do not think about what this fallback position might be when drafting your application (which means finding and understanding the closest prior art), you might

Inset: Prior Art

Prior art is often considered a nasty term to people responsible for building a patent portfolio. Two of the statutory requirements for getting a patent are that your invention must be novel (35 U.S.C. § 102) and non-obvious (35 U.S.C § 103). Prior art reading on your invention can therefore prevent you from obtaining an issued patent.

It is much better to know about the prior art during the application drafting stage than only after the application has been filed. Instruct your patent professionals as well as your R&D staff to take the time to thoroughly search and identify related inventions. Issued patents, published patent applications, articles in scientific and trade journals, abstracts of presentations at conferences, and other written materials are all part of the prior art. The more you know about what the prior art is, the better you will understand exactly what your invention is and why it should be patentable.

not be able to provide sufficient support for these fallback positions.

To provide a simple example, say your invention involves a process in which a substrate may be heated to between 100 and 150°C, but all of your examples are between 140 and 150°C. The examiner finds a prior art reference disclosing this step at 135°C and above. To get your application issued, you need to edit your claim to a temperate range of between 100 and 134°C. But because you never gave an example of performing this step within this range, the examiner may argue that you never enabled this temperature range, and thus you will not be able to claim this "fallback" position.

This problem commonly occurs in chemistry cases. Suppose you have developed a library of novel chemical compounds, and your invention is a new formulation of these compounds (ointments, gels, etc.). Suppose further that while you describe these formulations generally with the whole library of compounds, your examples—formulations you actually made—use only one of the compounds in your library. In such a case, while your patent application attempts to claim formulations to the entire library of compounds, there is a good chance that the examiner will reject the "genus" claim because an insufficient number of "species" of different individual compounds was given.[3] If this happens, you would only be able to claim formulations of that one compound.[4] You have a much stronger case for claiming formulations to the entire library if you can show examples of a representative number of species within the genus. Figuring out what this "representative" set is will be up to you and your attorney or agent.

2. Do Not Vaguely Describe "Non Core" Ideas

It is very common that when writing your first few patent applications, you have not only your main or "core" application in mind, but also

3. The law on genus and species claims is a bit more complicated than this, and is still unsettled in the courts. However, this practice is especially important in highly technical or "unpredictable" arts such as chemistry, biotechnology, and the like. It is better to err on the safe side and give as many examples as possible over the broadest range as possible.
4. Another way to overcome this rejection is to send post-filing data to the Examiner demonstrating that the disclosure was in fact enabled when it was filed. Beware that while this approach may get the patent issued, such declarations can also be used as fodder for invalidity attacks in litigation.

other ways in which your basic idea can be applied or adopted. It is very enticing to try to vaguely capture these ideas in your application without fully understanding the technology in which they might be applied. If these "non core" aspects of your invention are simple and can be easily practiced without "undue experimentation," then a court might enforce such claims. However, if these aspects of your invention require much experimentation or are in highly technical areas, claims to these applications might fail for want of sufficient enablement.

For example, let us say your invention is a rim of a bicycle wheel with a dimpled surface (think golf ball) that reduces wind resistance at high speeds. In principle, you could adapt this dimpled surface to anything that moves through air or water to reduce drag, save energy, increase speed, and so forth. So in your written description, after fully describing your slippery bicycle rims, you briefly mention that a dimpled surface can be applied to cars, airplanes, ship hulls, footballs, and Frisbee® discs. You confidently send your application in, waiting for phone calls from The Boeing Company and other major companies to license your idea.

What really happens? You are very unlikely to get claims issued to these other aspects of your core idea. Instead, you have written yourself into a corner: on one hand you have not sufficiently described how a dimpled surface would actually be applied on and reduce the drag of a Frisbee® disc or a ship hull, yet you have introduced these ideas into the public domain. This means that you may not be able to get claims on your first patent application, nor on later applications since, by introducing the idea literally, you increase the chances that this idea is "obvious" to anyone reading your application.

3. Define Terms Broadly and Provide Examples

Courts have repeatedly caused trouble for patent owners who have provided only a few examples of their invention. They have done so even after a landmark case[5] in which the court ruled that claims should not be limited to only the "disclosed embodiments" (the specific examples described in the specification).

In practical terms, it is better to provide several examples that

5. *See* Phillips v. AWH Corp., 415 F.3d 1303 (Fed. Cir. 2005) (en banc).

illustrate different aspects of your invention. For example, if your invention is a mechanical device with an internal support structure perpendicular to the main beam, check whether the support structure really needs to be "perpendicular." Can it be at another angle? If so, provide examples that illustrate a range of angles that will work. Similarly, when defining terms, be as broad as possible, but also provide narrower examples. For example, if you define the term "board" as a wooden support structure, you may be limited to wood. It is better to define the term broadly as a "support structure" made from a suitable material, for example, wood, plastic, metal, composite, etc.

4. Avoid "Profane" Terms and Phrases

The patent drafter can save a lot of trouble during litigation by avoiding certain damaging terms and phrases—the equivalent of four-letter words in patent law. Examples of "patent profanity" include the following:

- "The broadest aspects of this invention are …" and "This invention is …"
- "Very important," "critical," "essential," etc.
- "Preferred," and "most preferred"
- "This invention overcomes deficiencies in the prior art because …"

In litigation, the other side will use these phrases to limit the breadth of your claims. *And they will rip you to shreds!* Courts have held in numerous cases that when you write "my invention is X," it literally means your invention is "X" only. Forget "X*" or variants that you think are obvious or understood. For example, if you write that your "invention" is a method to electrostatically paint a car, your patent will not cover the same method to paint a bicycle or motorcycle. Do not even use the word "invention." It is much better to write that "in one aspect, we describe a method to electrostatically paint a car." Then you can describe other aspects, and your claims will be less likely to be limited. Other phrases such as "critical" or "preferred" will also be used to limit your invention as narrowly as possible, giving potential infringers room to freely practice what you thought would be covered by your patent. There is no reason to use these phrases when instead you can use phrases such as "in one aspect," "in certain embodiments," "by way of example," and so forth.

5. Write Claims for Potential Infringers

When thinking about blocking the competition, keep in mind who might try to practice your invention and write your claims accordingly. For example, let us say your invention is a method of cleaning a medical device *in situ* by rinsing the medical device with a cleaning solution comprising novel ingredient X. Claim 1 might read as follows:

> Claim 1. A method of cleaning a medical device, comprising (a) providing the medical device to a patient; and (b) rinsing the medical device with a cleaning solution comprising ingredient X.

While this claim might at first sound like it closely covers a potential infringer, by introducing limitation (a), however, you have limited your claim to cover only those people who actually provide the medical device to the patient. Suing individual doctors or nurses and collecting damages from them will be a nightmare. Moreover, what if someone sells a bottle of your ingredient X and does not provide the medical device? That party might not infringe your patent, since it did not practice all the limitations of your claim.

In summary, follow these practice pointers when drafting your patent applications:

- Describe narrower fallback positions in addition to your broadest idea.
- Stick with describing your main invention and make sure you fully enable it. Save filing patent applications on your related "non core" ideas until you have fully worked out the details.
- Define terms as broad as consistent with your invention,

and provide several specific examples to establish that breadth.

- Do not use profane terms in your application!
- Write claims with a specific potential infringer in mind, and make sure that potential infringer will be able to pay damages.

WHAT R&D NEEDS TO KNOW AND DO

The research and development department is often a great source of innovation and thus the driving force of new intellectual property. Assuming that most inventions will be coming out of R&D, your R&D staff can take several actions to improve the way IP is captured and developed. Here are a few thoughts.

A. WORKING WITH IP COUNSEL

The inventors will have the best idea of what the invention is, how it differs from the prior art, and what modifications can be made. Therefore, the inventors should work closely—even directly—with the agent or attorney writing the application. Appoint someone to be the main contact with the patent drafter. The lead inventor would be a good candidate for this role. Requiring the inventors to take ownership of the drafting process will help ensure that the attorney writes the strongest and broadest application possible.

B. WRITING THE EXPERIMENTAL SECTIONS

A patent is a type of contract with the government: in exchange for being able to exclude others from practicing your invention, you must teach the public how to make and use it. In this way, technical knowledge advances.

The courts have set a standard for the written description. Your application must describe the invention in such a manner that another person "with ordinary skill in the art" would be able to make and use the invention "without undue experimentation." These terms are those used by the courts, and will be interpreted taking into account the technology and complexity of your invention.

Examples of how to use your invention in different ways can be actual (ones that have been made) or "prophetic" (ones that have not yet been made but that nevertheless describe how one would use the invention in a certain way). Actual examples that have been made are properly described in the past tense, since they were actually made. Prophetic examples are written in the present tense, signaling they are illustrative but have not actually been made.

Instruct your R&D team to write the written description section in the same manner and level of detail as that of an article written in a peer-reviewed, trade, or similar journal or communication typical of that technology area. It must describe the invention in sufficient detail that someone else working in that technology area can read the application and make and use the invention with no more than a reasonable amount of head-scratching or failed initial attempts. Thus, the standard required for peer-reviewed journals is a good guide.

C. KEEPING GOOD NOTEBOOKS

Notebooks can be critical in establishing who invented the invention, and when. Engineers and scientists should be familiar with good notebook practices for their experimental work. When it comes to new inventions, the core aspects of the invention and a few illustrative examples should be described in a notebook that is signed and dated, and co-signed by a non-inventor. This non-inventor should be able to reasonably understand what is written and described in the notebook and ideally should be in a different research group as the inventor so as to remain impartial. Guidelines on keeping research notebooks are commonly available (*see, e.g.*, H.M. Kanare, *Writing the Laboratory Notebook*, American Chemical Society (1985).

D. INVENTION DISCLOSURE FORMS

Your patent attorney or agent may ask you for an "invention disclosure form" before starting to write the application. Such a form is usually used to formally disclose the invention of the inventors to the attorney or agent writing the application, and can be a useful tool in tracking important dates (such as dates of public disclosure), identifying the most relevant prior art documents, and making sure the invention is adequately disclosed. Your attorney or agent will provide the necessary forms to the inventors and explain how they should complete it.

WHEN AND HOW TO BRING WORK "IN HOUSE"

As your company grows, it will make sense at some point to bring some of your patent work "in house" by hiring your own patent attorney or agent. The economics of this decision is rather simple: if you are using outside counsel between about one-half to one day per week, a full-time internal position starts to look favorable.[6]

The duties of your first in-house attorney or agent will not only include substantive legal and technical aspects of writing and prosecuting patents, but will also include a significant amount of administrative tasks and giving general strategic advice. It is likely, however, that early-stage companies will not have document management software systems, and may not have much in the way of assistants or other administrative support.

Therefore, your first in-house IP person should—in addition to being technically savvy and a good writer—have a good ability to organize and manage files and deadlines with little to no administrative support. It is important to realize these requirements because most attorneys coming from big law firms may excel at performing legal tasks but, because they have administrative assistants, may not be comfortable performing many administrative duties. This ability is something you may want to determine during the interview stage. While writing the patent application is very important, missing a filing deadline can create big problems for an application.

Although your company may not have much in the way of administrative support or software tools, do not scrimp on your IP budget. Filing and maintaining patents is expensive. Patent offices strictly enforce deadlines for filing, paying annuity fees, and other requirements. The day you miss a deadline because no filing system was ever put into place is one day too late to realize the importance of adequately resourcing your IP department.

---

6. As a very rough calculation, a salary of $160,000 equates to about 6.7 hours per week at 48 weeks and $500/hour.

THOUGHTS ON HOW TO DEVELOP AN IP PORTFOLIO: FROM TACTICS TO STRATEGY

As you build your business and come up with new ideas, you will move from a position of having one or two patents to building a collection (or "portfolio") of patents and applications. While a detailed discussion of patent strategy is beyond the scope of this article, here are a few questions and considerations to keep in mind as you build out your portfolio.

A. KEEP YOUR BUSINESS GOALS IN MIND

What are your business goals? What is your competitive position? As patents make business sense only in supporting your commercial business objectives, answering these questions will be key in determining which patents you need to have and which patents will be nice to have. Focus on protecting your core ideas first. Then you can move on to protecting additional ideas, or working on ways to extent the patent life of your core commercial product.

B. MAKE SURE YOU HAVE FREEDOM TO OPERATE

Remember what a patent allows you to do: it does *not* allow you to make or use your invention—it only allows you to exclude others for practicing your claims. When developing your commercial product, conduct a search to make sure you are not infringing on the patents of others. If you think you are, you will need to consider either designing around them, or licensing them so that you will have "freedom to operate" in your chosen business area.

C. ARE YOU OR YOUR PARTNERS GOING INTERNATIONAL?

Are you looking at selling your product in international markets? When? What about any current or prospective business partners? Answering these questions can be difficult but must be done up front: once your patent application publishes in the U.S., it is too late to file internationally. While filing, prosecuting, and maintaining pending and issued patents abroad can become very expensive, some partners will require you to have international applications before dealing with you. Know what is standard in your industry.

Different countries can be thought of as being in different "tiers" according to the size of their markets and the degree to which they enforce patent laws. To help manage costs, consider filing only your "core" patent applications in as many tiers as possible, and file your other applications in higher-tier countries. Tier one countries—the most important—typically include the U.S., Europe, Canada, and Japan. Tier two countries typically include Australia, Brazil, China, India, Mexico, and South Korea. Finally, tier three counties include Hong Kong, Singapore, South Africa, Argentina, Taiwan, and others. While there are no hard-and-fast rules in determining these tiers, and developing countries such as India and China are improving their patent policies, the tiers above can be useful—among other considerations—in devising an international patent strategy.

CONCLUSION

Intellectual property is an important aspect of most technology-based businesses. From writing and filing your first patent application to building and managing a patent portfolio, patent law can be a very complex, detail-oriented, and nuanced aspect of your business. Armed with the key concepts, however, business executives can manage their patent counsel effectively and ensure that patents support their business objectives.

CHAPTER 3

Present Patenting for Future Business Strategies

Oscar J. Llorin, Esq.

Oscar J. Llorin is a registered patent attorney and has been in practice before the U.S. Patent and Trademark Office (PTO) since 2001. His patent prosecution experience includes drafting and prosecuting patent applications in biotechnology, including the fields of DNA probe and antibody-based diagnostics.

During law school, Mr. Llorin was a judicial intern for the Honorable Randall R. Rader, Circuit Judge (now Chief Judge), at the U.S. Court of Appeals for the Federal Circuit and a judicial intern for Administrative Patent Judges at the PTO's Board of Patent Appeals and Interferences. Mr. Llorin earned his J.D. degree from the University of Maryland School of Law and received the national Order of Barristers honor.

Prior to law school, Mr. Llorin earned a B.S. degree in biology, with an emphasis in microbiology, from Virginia Tech and an M.S. degree in applied molecular biology from the University of Maryland. Mr. Llorin developed nucleic acid amplification assays for the diagnosis of infectious diseases at BD Diagnostic Systems and is an inventor of several patented inventions. Mr. Llorin can be contacted at ollorin@edwardslaw.pro.

A business executive who navigates his company through a competitive marketplace should be aware of the potential strategic uses of a patent. When used effectively, a patent can be worth much more than a plaque on a wall. Essentially, a patent gives one the legal right to exclude others from making, using or selling an invention. The patent owner can exercise this exclusionary right strategically by using the patent as either a sword or shield in the marketplace against competitors. When the patent owner owns numerous patents, each covering different aspects of the invention and varying in scope, greater leverage in the marketplace is possible.

For example, a patent portfolio can serve marketing purposes, help generate quick revenue, or serve long-term strategic needs. In order to implement these strategies, the executive must begin preparations as early as when the patent application is first drafted and subsequently prosecuted before the U.S. Patent and Trademark Office (PTO). Strategic patenting thus requires an understanding of (1) the specific components of a patent, (2) how these components can be drafted to implement business needs, and how to effectuate long-term strategies (3) through the prosecution of pending patents and (4) use of post-issuance tools at the PTO.

KEY PATENT COMPONENTS

A. THE CLAIMS: METES AND BOUNDS OF PROPERTY

Good patent drafting starts with drafting the section that appears at the very end of every patent issued and published by the PTO: the claims. The PTO will only allow claims if it determines that the claims described a process, machine, article of manufacture or composition of matter. Moreover, the claims must be new and non-obvious. To determine novelty and non-obviousness, the PTO searches the prior art already present in the public domain.

Patent drafters may start with the claims because the boundaries or scope of property rights are determined here. In other words, the claims give the public notice as to what it cannot make, use or sell. The claims do so by using language to demarcate the "metes and bounds" of the owner's inventive property, like a fence on a map outlining the boundaries of land covered by a deed. In a claim, limitations or elements detail features of the inventive concept. The limitations or elements are analogous to fence posts connecting the boundary lines of land on a map. If a patent owner wants to exclude an infringer, e.g., a competitor who makes, uses or sells what is covered by a claim without the patent owner's permission, the patent owner must prove that the accused infringer's goods or actions meet every limitation or element of the claim. Patent claims can be broad or narrow; the pros and cons of each are discussed below.

Broad claims will contain relatively few limitations or elements and are generally shorter in length. The fewer the limitations, the easier it is to prove infringement. Broad claims provide the greatest patent protection by covering a larger range of situations or valuable products. In other words, broad claims increase the amount of land covered by the patent owner's fence. Broad claims, however, are harder to patent because the breadth of the claim will oftentimes include features already present in the prior art. For patent prosecution, this will mean that it will likely take longer for the patent to issue. The more prior art references the PTO identifies to support its references, the longer it will take for your patent attorney or agent to overcome the PTO's rejections. Broad claims are also harder to enforce because they increase the likelihood that your claim can be invalidated during litigation or reexamination proceedings. Thus, it will be easier for your competitor to find prior art to invalidate your patent.

On the other hand, narrow claims contain numerous limitations, with greater detail, more specific language and are generally longer. In other words, narrow claims contain numerous landmarks demarcating the inventive property. These claims provide minimum protection, because they will not cover as many situations or valuable products as a broad claim. Consequently, fewer competitors are excluded from making, using or selling your business's invention. Accused infringers are delighted when the asserted patent claim is long. This increases the chances that they do not meet every limitation of the claim, thereby improving the chances of mounting a successful non-infringement defense. Filing a patent application with narrow claims can, however, provide some advantages. Patent attorneys and agents will likely have a shorter time prosecuting narrow claims in the PTO because there is less chance that the PTO can cite prior art that meets every limitation or element of the claim.

B. THE SPECIFICATION: SUPPORT FOR CLAIMS

To determine the meaning of the claims, the PTO (during patent prosecution or a reexamination) and federal courts (during infringement proceedings) will look to the patent's specification. This is necessary because nearly everything included in the patent prior to the claims is drafted to support the claims. The specification, also called the disclosure, provides a written description of the invention. At the bare minimum, it must satisfy statutory requirements for patentability. Depending on how it is drafted, specific parts of the disclosure can determine the scope and breadth of the claims.

A well written specification provides a persuasive story of how the inventors were able to solve a problem or address a previously unmet need with a solution or an improvement. The specification may begin with the Background of the Invention, which describes limitations and problems in the prior art. Depending on the patent owner's needs, the patent drafter can cite few or many prior art references in this section. If the patent applicant is very familiar with the prior art, discussing it in the Background of the Invention might make sense. On the one hand, describing the differences up front may be more persuasive, but on the other hand, it is easier for these comments to be used against the patent owner later in patent litigation or reexamination proceedings.

This eventually leads into the Detailed Description of the Invention, which describes the inventive concept's solutions and improvements applied to the problems or shortcomings described earlier in the prior art. The Detailed Description of the Invention includes layered explanations of the invention. First, a general explanation of the invention provides support for the broadest view of the claimed inventive concept. Second, preferred embodiments of the invention may be included that are more limited than the broadest inventive concept. These preferred embodiments may support narrower claims. The patent drafter may also include figures and the definitions of key terms found in the claims. Specific examples of how the claimed invention is practiced may be included, although these are not required. If included, these examples describe the practice of one or more specific embodiments of the invention. Typically the longest section of the patent, the Detailed Description of the Invention should be drafted with sufficient disclosure to satisfy statutory requirements for describing and enabling the invention. The language used to describe features of the invention in this section should match the claims language to adequately protect all embodiments of the invention.

How each section of the claims and specification is drafted will impact whether business strategies, implemented under the legal instrument of the patent, are successful. Below are some examples of how patent drafting impacts future strategies in the marketplace.

EFFECTIVE PATENT DRAFTING TO IMPLEMENT BUSINESS STRATEGIES

A. CLAIMING TO EXCLUDE OTHERS: YOUR PATENT AS A SWORD

When using the patent as a sword, a business can strategically exclude its competitors from any commercial activities in the marketplace, as long as the PTO grants the business a patent with claims which cover the competitor's potential commercial activities, e.g., selling a valuable product. At the very least, the competitor need not have actual knowledge of the excluding claim. Once an excluding claim is granted by the PTO, the competitor is deemed to have constructive knowledge and the patent owner can seek damages and get injunctive relief to stop the competitor's infringing activities. Better yet, if the competitor demonstrates objective recklessness with regard to an issued claim and chooses willfully to go ahead with the infringing activities, patent law affords the patent owner up to treble damages and payment of attorney's fees.

For these reasons, it is very important for the executive to work with her patent drafter up front to ensure that the excluding claim in an application is drafted to support three goals. First, the executive must ensure that the claims language actually covers the future commercial product. Second, the executive must make sure that the claims include language that targets literal infringement. Third, the executive must keep in mind predictable variations of the product. In other words, the claims should also cover products reverse engineered by the businesses competitors. To do this, the executive should encourage inventors to think of all foreseeable equivalents. Failure to include an equivalent in the specification or the claims may later prevent the patent owner from successfully arguing that the competitor infringes that equivalent.

The more excluding claims the better. Inclusion of a large number of claims provides a problem for competitors, if the claims vary in their scope of coverage through different wording. Excluding claims should also include language which buys the executive more bang for his buck. Ideally, these claims should ensnare an infringer with deep pockets by covering products with the highest value, thereby enhancing damages. For example, if the claim is directed to a touch screen, inclusion of a mobile phone, a computer, an ATM machine, etc. as limitations which incorporate the touch screen will expand the list of defendants from which an executive can seek damages.

B. DEFENSIVE CLAIMS IN YOUR PATENT SHIELD.

As a shield, the patent owner can use the patent to defend herself from a competitor's lawsuit. For example, if the business is sued by a competitor for patent infringement, the defendant can use her own patents to file a counterclaim against the competitor. Use of a patent or ideally a patent portfolio defensively in this way can lead to a settlement agreement between the parties to cross-license each party's patents. If the defensive patent was first drafted with numerous broad claims, the issued patent should cover much more than the actual invention and is likely to overlap claims found in a competitor's patent portfolio. This will increase the odds of cross-licensing, especially in emerging technologies with patent thickets.

C. GETTING CLAIMS ISSUED TO RAISE REVENUE.

As mentioned above, narrow claims contain all of the features of the main invention that distinguish it from the closest prior art. The PTO will oftentimes allow the narrow claims and reject the broader claims in the patent application.

A start-up company in need of revenue can improve its marketability and attractiveness to investors by generating a patent portfolio. Because time is of the essence in this scenario, the patent applicant can draft narrow claims drawn to the nascent company's invention(s), ignoring, for the time being, capturing any infringing activities practiced by licensees with broad claims. The specification in this scenario need only contain the minimum amount of disclosure to support the narrow claims. Filing a quick, dirty, and relatively cheaper patent application in this manner facilitates faster issuance. Keep in mind, however, these patent applications may not contain enough disclosure to support broader "children" patents, as discussed further below.

An established company may desire generating additional revenue quickly through commercial sale of a first generation product in a breakthrough technology. With additional capital, the company can develop an extensive patent portfolio by filing a comprehensive patent application as the initial foundation. This patent application should contain a wide range of embodiments and working examples in the specification. Additionally, the applied claims should vary in breadth and scope. During prosecution, the patent attorney should cancel broad claims in favor of obtaining a patent that covers the specific first generation product with narrow claims. As described below, broad claims can be included in subsequent continuation applications. This can ensure the development of a patent portfolio for next generation products without delaying the initial product launch.

If the goal of the business is to raise revenue through licensing, the executive must ensure that claims cover more than the business's product to be sold. In this scenario, the patent application must include claims that specifically cover what the prospective licensee is doing. The executive should also consider including claims covering the product in different fields or environments. Different product uses should be addressed as well in separate claims. For example, if the invention is for an isolated nucleic acid sequence composition associated with ovarian cancer, then additional claims could cover use of that sequence in diagnostic assays or for therapeutic uses in other organs. The executive must make sure that the patent drafter includes adequate support for these additional claims in the specification.

In sum, whether the business owner elects to strategically use the patent as a sword or a shield, or to generate revenue, either choice requires careful attention to how the patent application is prepared and prosecuted before the PTO.

PROSECUTION OF PENDING APPLICATIONS TO IMPLEMENT LONG-TERM GOALS

If the PTO is convinced that the invention as claimed and described in the specification meets statutory requirements, it will issue a notice of allowance. The executive should keep in mind that the PTO allows a business to file additional patent applications before the PTO issues the original "parent" patent or the "parent" application is abandoned. These additional "child" patent applications may contain claims which differ in scope from the claims the PTO has allowed in the parent. The types of child patent applications that can be filed before the parent issues include continuations, divisionals and continuations-in-part. A business can use each type, depending on the situation, as a strategic tool to gain a competitive edge in the marketplace.

For example investors are generally attracted to start-ups company with an established patent portfolio. Investors, however, may steer away from start-ups who only own patents with broad claims at risk of invalidation in PTO reexamination proceedings or litigation disputes in federal courts.

If the business has filed a pending foundational application, with both narrow and broad claims, the broad claims can be deleted from an initial parent patent application in favor of relatively quick PTO issuance of the narrow claims. This may facilitate investor interests in the short term. In the long term, the broad claims can be included in a continuation application, which, of course, must be supported by the original specification. Filing a child patent in this manner can result in the subsequent issue of a continuation patent with claims drawn to broader aspects of the invention.

Keep in mind this can only occur if the pending application contains enough disclosure to support both narrow and broad claims. If it is drafted as a foundation for a future portfolio, it should be very detailed, disclose as many embodiments as possible and it may even include definitions for claim terminology. This type of patent application is more expensive to draft, but worth the costs because it can provide the expansion of the intellectual property estate via child patents. On the other hand, a cheaper, quick and dirty application might not contain a sufficient disclosure to support continuations.

What if the executive encounters a competitor who has "designed around" a patent? If the patent owner has kept alive a pending continuation application, new claims can be added that are specifically drafted to cover the competitor's infringing activity. If the continuation patent issues, the executive can seek a license or file suit against the competitor. Used in this manner, pending applications allow the executive to adapt the business's inventive concept to future developments and technological advances.

Pending application practice provides a useful tool for businesses to expand their patent protection. Executives should revisit their pending applications on a regular basis to ascertain opportunities for continued filings. If, however, a patent is no longer pending and has already issued, the PTO provides additional tools to further an executives strategic needs.

POST-ISSUANCE TOOLS TO IMPLEMENT LONG TERM GOALS

Suppose a business is the new owner of a recently issued patent in a breakthrough technology. An executive of that business, conducting due diligence on his competitor, learns that the adversary has a pending patent application with claims directed to the same invention. If the executive's patent issued within one year of the pending application's filing date, the PTO will require both parties to appear in an interference proceeding to determine who is awarded patent rights. U.S. patent law currently awards patent rights to the first inventor to conceive of the invention and diligently reduce it to practice, although proposed legislation would award patent rights to the first to file the patent application. If the executive's business prevails in this proceeding it can commercialize the invention unhindered by his competitor. However, the business may lose its patent rights outright if it fails to win in this complex and costly proceeding. Executives should note that the PTO will mandate pre-grant interference proceedings if it encounters two patent applications from different inventors covering the same invention, although pending legislation may change this.

If the business has recently been awarded a patent, but an executive determines that the granted patent is, unfortunately, defective in whole or in part, invalid, or just does not provide the desired scope of claims protection, then she should consider applying for a reissue patent. Prudence is necessary here. It will be better for the patent owner to cure an invalid patent and retain patent rights, rather than suffer the invalidation of the patent by a competitor's successful litigation or reexamination. To take advantage of this post-issuance tool, the original specification must support the newly desired claim scope. Also, the original patent must be surrendered after an additional PTO examination. Broadening reissues are only allowed within two years of patent issuance. On the other hand, narrowing reissues can take place anytime between patent issuance and expiration.

Finally, if an executive encounters a previously unknown prior art reference which could jeopardize the patentability of his granted patent, the executive should consider a PTO reexamination. This would strategically offset the possibility of a competitor using that prior art reference to invalidate the patent in a future lawsuit. Claims may be reduced in scope by reexamination, but this would be preferable to having your patent invalidated later in court. Reexaminations can be filed anonymously. Executives can even file them offensively if a substantial new question of patentability exists as to the claims in a competitor's patent.

CONCLUSION

Outlined above are only some of the important patent drafting, prosecution and post-issuance tools an executive can strategically use to maximize the value of a business' patent rights. Used effectively, the patent right can provide substantial advantages to the detriment of an executive's competitors for decades in the marketplace. Because this article is intended as an introduction to these strategic tools, executives should work closely with their patent attorneys and agents when procuring patent rights. Competent registered patent practitioners pay detailed attention to the complex and arcane rules of patent law to give your business a competitive advantage.

CHAPTER 4

The Tension Between Academic Authorship and Patent Inventorship

John A. Evans

Dr. Evans is an associate in the Intellectual Property Group in Pillsbury Winthrop Shaw Pittman's Washington, DC office. His practice primarily focuses on Hatch-Waxman litigation and patent prosecution in the biotech and pharmaceutical industries. Dr. Evans' practice focuses on the fields of chemistry and biotechnology, including organic chemistry, biochemistry, recombinant DNA technology, monoclonal and polyclonal antibodies, enzymes, chelating agents, medical diagnostics, imaging, pharmaceuticals, medical devices, implants, and semiconductor devices. He has extensive experience in patent application preparation and all aspects of U.S. and foreign patent prosecution. He provides counseling to clients on infringement and validity issues and the market clearance of products. Dr. Evans can be contacted at evans.jae@verizon.net.

One lurking danger for the academic new to the world of patenting is the difference in how authors are named. Just as those outside of academia may not appreciate the distinction between being a first author and a second author, there is far more at stake in naming patent inventors than would first appear. This chapter walks through a typical situations to point out the pitfalls.

The interplay between scientific authorship and invention may be seen in a hypothetical situation. The Aids Research Group of Venture Pharma Corp. collaborated with Professor Tenured's lab at State U. to develop and test anti-retroviral drugs. The collaboration was carried out pursuant to a research contract under which Venture Pharma paid State U. to have its personnel test compounds supplied by Venture Pharma. Dr. Lead Scientist of Venture Pharma provided a series of potential anti-retroviral compounds to Professor Tenured for testing in his novel assay that used live HIV in human T-cell clones. Dr. Scientist had synthesized AZZT and had shown that compound to be active against mouse retroviruses. Dr. Scientist supplied AZZT and several other compounds to Prof. Tenured. The compounds were coded so that Prof. Tenured did not know their real identities.

Prof. Tenured assigned his first-year graduate student, Sam, to perform the T-cell assays, the details of which Dr. Tenured has worked out previously. Originally, the T-cell line had been a gift of the department chair and the cell line was tended and propagated by a skilled technician. During preliminary trials, Sam learned that the AZZT was incompatible with the medium used to assay the T-cells. Mike, a post-doctoral fellow in Prof. Tenured's lab, worked with Sam, each suggesting improvements to the assay medium to make it compatible with both T-cells and AZZT.

Upon finding that AZZT had remarkable anti-retroviral effects in the human T-cell line, Dr Tenured ordered Mike to draft a manuscript for publication in a leading scientific journal. Tenured determined that he should be the first-named author, the department chair the last-named author, Mike the second author, and the technician should be the third-named author. Tenured concluded that Sam was as yet too junior to be awarded an authorship of such a potentially important paper. Sam pouted, but he had no other more effective means of redress.

Dr. Scientist drafted an invention disclosure which he forwarded to Venture Pharma's patent counsel. Based on that disclosure, patent counsel drafted a patent application and an assignment document. The application contained several AZZT related claims:

- Claims to the AZZT compound;
- Claims to methods of making the AZZT compound;
- Claims to pharmaceutical compositions containing the AZZT compound;
- Claims to methods of assaying an anti-retroviral

activity using T-cells; and

- Claims to methods of treating a human infected by HIV comprising administering an effective amount of AZZT.

Under his employment contract, the rights to all inventions made by Dr. Scientist made on company time and using company resources were to be assigned to Venture Pharma. Venture Pharma's patent counsel filed the patent application and the assignment in the U.S. Patent and Trademark Office (USPTO). Venture Pharma licensed the rights to any patent that might issue from their AZZT application to MegaPharma. The license called for Venture Pharma to receive initial payments during the remaining research phase of the AZZT project and to receive royalties on any sales of AZZT after it received FDA approval as a drug.

Shortly after filing their patent application, Venture Pharma gave Prof. Tenured a copy of the application as a courtesy to help him in preparing a grant application to be submitted to the National Institutes of Health (NIH). The officials of State U. were furious that Venture Pharma drafted an application on their own, and argued that the authors of the journal article should be named inventors on the AZZT application and that State U. was entitled to share the license payments and any future royalties.

The authorship of a scientific journal article is subject to custom and political considerations. Often, the first-named author is the person who drafts the manuscript and the head of the laboratory is the last-named author. According to this custom, Mike might have expected to be the first author and Tenured would be the last-named author. However, where it is expected that the paper might become significant in the field, often the head of the laboratory will "pull rank" and name himself the first author. Sometimes authors are listed alphabetically. Often, it is politically expedient to name as authors, persons, such as the department chair, who have made only minor contributions to the actual project. Persons, such as the technician, who have made little, or no, intellectual contribution, but who have provided valuable skilled labor, are often rewarded with authorship. Low-ranking persons, such as Sam, may get squeezed out.

Unlike authorship, patent inventorship is governed by law—so you are not allowed to pull rank. Failure to name a person who is lawfully an inventor, as well as naming as an inventor someone who does not qualify legally may have serious adverse consequences. These consequences may include invalidation of the patent and loss of royalty rights. Such challenges to the validity of the patent often arise in circumstances where there are multiple, academic authors.

The Patent Statute provides that only the inventor may apply for a patent[1] and only after declaring that she believes herself to be the original and first inventor.[2] The authors of the scientific paper claim that they are joint inventors with Dr. Scientist. People may be joint inventors even though they do not physically work on the invention together or at the same time, and even though each does not make the same type or amount of contribution.[3] Although the statute does not set forth the minimum quality or quantity of contribution required for joint inventorship,[4] each co-inventor owns a *pro rata* undivided interest in the entire patent, no matter what their respective contributions.[5] Because each inventor owns a property interest in the patent, there is often incentive to challenge the designated inventive entity to name oneself an inventor. In addition, litigation opponents also have an incentive to challenge the inventorship of the patent to render the patent invalid.

Under the law, patents are personal property;[6] property rights, including ownership, attach to patents as a whole, not just to individual claims. Moreover, Section 262 continues to speak of "joint owners of a patent," not joint owners of a claim. Thus, a joint inventor to even one claim owns an undivided interest in the entire patent.

In the above scenario, who is a lawful inventor? "The exact parameters of what constitutes joint inventorship are quite difficult to define. It is one of the muddiest concepts in the muddy metaphys-

1. 35 U.S.C. § 111 (1987).
2. 35 U.S.C. § 115 (1987).
3. *Borroughs Wellcome Co v. Barr Labs, Inc.*, 40 F.3d 1223, 1227 (Fed. Cir. 1994) (*citing* 35 U.S.C. § 116).
4. *Borroughs Wellcome Co v. Barr Labs, Inc.*, 40 F.3d 1223, 1227 (Fed. Cir. 1994).
5. *Ethicon, Inc. v. U.S. Surgical Corp.*, 135 F.3d 1456, 1465 (Fed. Cir. 1998).
6. *Ethicon, Inc. v. U.S. Surgical Corp.*, 135 F.3d 1456, 1465-1466 (Fed. Cir. 1998) (*citing* 35 U.S.C. § 261, "patents shall have the attributes of personal property").

ics of the patent law."[7] The first step to determine inventorship is to review the claims. Inventorship is determined on a claim by claim basis, therefore, each claim must be analyzed separately.

The first set of claims, to the AZZT compound, were invented by Dr. Scientist. "Conception is the touchstone of inventorship, the completion of the mental part of invention. It is "the formation in the mind of the inventor, of a definite and permanent idea of the complete and operative invention, as it is hereafter to be applied in practice."[8] "Conception of a chemical substance, [such as the hypothetical AZZT] includes knowledge of both the specific chemical structure of the compound and an operative method of making it."[9] A person is an inventor if she conceives the idea of a claim. In our hypothetical, prior to contacting his collaborators at State U., Dr. Scientist had synthesized AZZT and, based on its effects against mouse retrovirus, conceived that it might be effective against human retrovirals. Prof. Tenured argues that without the results obtained using his T-cell assay that Dr. Scientist did not have a reasonable expectation that AZZT would be effective against human retrovirals. However, "an inventor's belief that his invention will work or his reasons for choosing a particular approach are irrelevant to conception."[10] The inventor does not need to know that his invention will work for conception to be complete. He needs "only show that he had the idea; the discovery that an invention actually works is part of its reduction to practice."[11]

Because conception "is a mental act, courts require corroborating evidence of a contemporaneous disclosure that would enable one skilled in the art to make the invention."[12] Here we will assume that Dr. Scientist followed industry-standard laboratory notebook prac-

7. *Mueller Brass Co. v. Reading Indust.*, 352 F. Supp. 1357, 1372 (E.D. Pa. 1982), aff'd. 487 F.2d 1395 (3rd Cir. 1972).
8. *Burroughs Wellcome Co v. Barr Labs, Inc.*, 40 F.3d 1223, 1227-8 (Fed. Cir. 1994) (citations omitted).
9. *Burroughs Wellcome Co v. Barr Labs, Inc.*, 40 F.3d 1223, 1229 (Fed. Cir. 1994) (citations omitted).
10. *Burroughs Wellcome Co v. Barr Labs, Inc.*, 40 F.3d 1223, 1228 (Fed. Cir. 1994) (citations omitted).
11. *Burroughs Wellcome Co v. Barr Labs, Inc.*, 40 F.3d 1223, 1228 (Fed. Cir. 1994) (citations omitted).
12. *Burroughs Wellcome Co v. Barr Labs, Inc.*, 40 F.3d 1223, 1228 (Fed. Cir. 1994) (citations omitted).

tice, including having each entry dated and witnessed.

The methods of assaying AZZT were developed pursuant to the research contract for which Venture Pharma paid State U. Therefore, Venture Pharma argued that Dr. Scientist was also the inventor of these claims and the personnel of Prof. Tenured's lab were merely contractors. State U. argued that the authors of the journal article should be named as co-inventors because they had contributed to the conception of the assay method claim. An agreement was sought whereby the four journal authors would be named co-inventors with Dr. Scientist and they would agree to assign their rights to Venture Pharma.

We recall that some of the State U. personnel were included or excluded from the authorship roster for political considerations. However, applying such considerations to a patent application may invalidate any subsequent patent.[13] Venture Pharma's patent counsel determined that of the journal authors, only Mike, the post-doctoral researcher, had contributed to conceiving the modified assay. The department chair had supplied the critical T-cell line, but merely supplying materials is not an inventive act. The inventor "may use the services, ideas and aid of others in the process of perfecting his invention without losing his right to a patent."[14] Moreover, neither Professor Tenured nor the technician who cultured the T-cells were inventors. A modified agreement was reached naming Scientist and Mike as inventors. Mike assigned his patent rights to Venture Pharma as had Scientist done prior.

In due course, the USPTO awarded the '999 patent to inventors Scientist and Mike. Subsequently, MegaPharma learned that UltraPharma was infringing the patent by assaying the effects of anti-retrovirals using human T-cells, as claimed. MegaPharma instituted a patent infringement action. As luck would have it, and while the suit was still pending, UltraPharma became aware of Sam, the disgruntled graduate student. UltraPharm obtained from Sam a "retroactive license" to practice Sam's T-cell assay related inventions. Under

13. 35 U.S.C. § 256 (1982) ("The error of omitting inventors or naming persons who are not inventors shall not invalidate the patent in which such error occurred if it can be corrected as provided in this section").

14. *Ethicon, Inc. v. U.S. Surgical Corp.*, 135 F.3d 1456, 1460 (Fed. Cir. 1998) (citations omitted).

the license, Sam agreed to assist UltraPharm in any suit regarding the '999 patent. UltraPharm agreed to pay Sam contingent on its ultimate ability to continue to practice and market the invention. Then, UltraPharm moved to correct inventorship of the '999 under 35 U.S.C. § 256, claiming that Sam was a co-inventor of claims to methods of assaying retrovirals using human T-cells.

Following an extensive hearing, the district court granted UltraPharm's motion, finding that Sam had contributed to the conception of the T-cell assay claims. UltraPharm next successfully moved for dismissal of the infringement suit, arguing that Sam, as a joint owner of the patent, had granted it a valid license under the patent. By its terms, the license purported to grant rights to use the patent extending retroactively back to its issuance.[15] Thus, conflating the customs and practice of scientific authorship with the legal requirements for inventorship, may have dire adverse consequences for any patent that may be granted.

15. This fact pattern was taken from *Ethicon, Inc. v. U.S. Surgical Corp.*, 135 F.3d 1456 (Fed. Cir. 1998).

CHAPTER 5

Obtaining Patents and Trade Secrets for Health Technology Companies

Kevin W. McCabe

Kevin W. McCabe is a Director at Sterne, Kessler, Goldstein & Fox, P.L.L.C. He specializes in intellectual property issues for health technology companies including medical device companies and companies delivering health-saving medicines to patients. The views expressed herein are the author's own and should not be attributed to either his firm or his past or present clients. Mr. McCabe can be contacted at kmccabe@skgf.com.

The value of intellectual property (IP) to the start-up or early stage company, especially those companies focused in the areas of biotechnology and medical technologies (collectively, the health technologies), cannot be underestimated. Although all industry segments rely upon IP, health technology companies are particularly reliant upon IP. Therefore, it is critically important for health technology companies to understand the various aspects of IP, what each form of IP protection provides, and the limitations associated with each IP form.

WHY IS IP SO IMPORTANT TO HEALTH TECHNOLOGY COMPANIES?

To understand the importance health technology companies place on intellectual property, it is useful to contrast health technology companies to its polar opposite, the electronics industry.

NUMBER OF PATENTS PROTECTING IMPORTANT COMMERCIAL PRODUCTS

In comparison to a cell phone, for example, which may be protected by tens of thousands of patents, a commercially important health technology product is typically covered by only a handful of patents. Moreover, in the electronics industry, it is relatively rare that a patent covers the entire product, whereas in the health technology industry patents more frequently cover most, if not all, aspects of the commercial product. Thus, reverting back to the cell phone example with its tens of thousands of patents, those patents cover different aspects of the product and collectively cover the phone itself. Whereas in the case of a pharmaceutical drug or a recombinant antibody the most valuable claims, and the ones most frequently sought, cover the whole of that pharmaceutical product or that recombinant antibody.

INVESTMENT TIMEFRAMES

Investment timeframes in the health technology industry often span a decade or more before a product can be brought to the market. First, one has to discover the product. That product invariably undergoes rigorous research to confirm that the product is efficacious. Then, the product is subject to clinical studies to further prove that the product is safe. Thus, it is not uncommon that it can take seven years for a product to go from discovery to commercial launch and periods of ten or more years of research and development are not atypical.

OTHER BARRIERS TO ENTRY

Other than regulatory barriers, patents provide one of the few barriers to entry to prevent others from copying a health technology product. Indeed, most health technology products are easily replicated once the invention has been made.

WHAT IS IP?

There are four basic forms of IP: patents, trademarks, copyrights, and trade secrets. Most important to the health technology company are the first and last forms of IP, however all forms of IP can be used to protect a health technology company's product.

PATENTS

Patents protect ideas. They are like deeds of property that define the property right of the owner. It is a frequent misunderstanding that patents provide the patent owner the freedom to make and use the patented invention. That is not the case.

By way of quick illustration, two companies are pursuing a therapeutic treatment for Alzheimer's. The first company discovers, and patents, the general concept that Alzheimer's can be treated using an antibody to a particular protein. The second company discovers, and patents, the use of a particular antibody against a particular domain on that protein. The first company after continued study arrives at the same domain and also seeks regulatory approval to a modified form, e.g., a more stable antibody, of an antibody to the same domain. In this example, the first company could not market its product because of the second company's patent. Similarly, the second company could not market its product because of the first company's patent.

Like all property, not every patent has equal worth. Some patents, like some deeds of property, are practically worthless, while others are extremely valuable. The value of a patent is measured by its scope as well as the particular technology that it covers. Typically, the most valuable patents are those that have extreme breadth and cover important commercial products. Typically, however, patents are obtained well before the commercial value of a product has been realized. Indeed, most patents are worthless, not because they do not describe and claim interesting and potentially very useful inventions. Rather, most patents have very little worth because they protect ideas that have no commercial application.

The breadth of a patent is defined by the claims, which are numbered paragraphs that appear at the end of a patent. These paragraphs define what the patent covers, i.e., what the patent owner can exclude others from making, using, selling, or importing, without permission from the patent owner.

WHAT IT TAKES TO OBTAIN A PATENT

To obtain a patent, the invention must have utility and it must also be novel, non-obvious, and described in a way that allows another person skilled in that particular field to make and use the invention. The governmental agency that undertakes the review of a patent application is the U.S. Patent and Trademark Office (USPTO). This agency is charged with reviewing a patent application to confirm that the inventor has satisfied the statutory criteria for obtaining a patent.

PATENT ELIGIBLE SUBJECT MATTER AND UTILITY

As a threshold consideration, the USPTO determines whether the claimed invention encompasses patent eligible subject matter. According to the patent statute, "[w]hoever invents or discovers any new and useful process, machine, manufacture, or composition of matter, or any new and useful improvement therefor, may obtain a patent therefor, subject to the conditions and requirements of this title."[1] Under this broad definition, laws of nature, mental processes, natural phenomena, mathematical formulae are not patentable. However, what is patentable is "anything under the sun that is made by man."[2] In the *Chakrabarty* case, the Supreme Court definitively settled the question whether a recombinant bacterial cell line was patentable. In short, such cell lines were patentable because human hands were involved in introducing the recombinant genes into the bacteria. Largely ever since then, it has been presumed that almost all inventions in the health technology industry fall within the broad definition of patent eligible subject matter. Indeed, the *Chakrabarty* decision has been widely cited as paving the way for much of the patenting by biotechnology companies in the decades since that decision.

Utility is a separate consideration from patent eligibility. Utility is shown where the claimed invention can achieve at least some useful effect. However, where the claimed invention is "totally incapable of achieving a useful result," the claimed invention lacks utili-

1. 35 U.S.C. § 101
2. *Diamond v. Chakrabarty*, 447 U.S. 303 (1980)

ty.[3] This is generally a low threshold. But nonetheless, it is not an idle doctrine. For example, in regards to ESTs, expressed sequence tags, the Federal Circuit held that ESTs lacked substantial utility unless the EST was tied to a particular function.[4] For most inventions, it is easy to establish function. This is why utility is a low threshold.

NOVELTY

A patent is novel when no one has previously described, used, or made the claimed invention. Novelty is destroyed where all of the limitations of the claimed invention are found in the prior art. However, the absence of any claim limitation in the prior art means that the claimed invention is novel.

In the U.S., a patent application must be filed within one year of its publication, public use or public disclosure by the inventor. Otherwise, the inventor's own work can be used to anticipate the later filed patent application. Europe and most other parts of the world follow a stricter absolute novelty test, meaning that the patent application must be filed before disclosure or the rights to a patent will be lost. This is one reason why companies and universities are so adamant about having patent applications on file before papers are published or the scientist is permitted to talk about the invention, for example, at a symposium.

Normally, this is easy to monitor and mistakes are not routinely made. However, there is one category of publications where mistakes are occasionally made. With respect to university-based inventions, universities must be vigilant about filing patent applications before doctoral theses publish. A litmus test for whether a publication is prior art is whether it is publicly available. Generally, if the publication has been cataloged and indexed, it is publicly available. Because most universities catalog and index doctoral theses in their libraries, the date that these theses have been cataloged and indexed in the library becomes their "publication" date.

The public use bar is another area where valuable patent rights are occasionally lost. Because of the public nature of universities, some inventions have been lost because of the public's observation

3. *Brooktree Corp. v. Advanced Micro Devices, Inc.*, 977 F.2d 1555 (Fed. Cir. 1992)
4. *In re Fisher*, 421 F.3d 1365 (Fed. Cir. 2005)

of the invention in practice before the critical date of the patent application. The classic example of this occurred at the National Institutes of Health, where a particular invention was being tested in a public room because that was where the equipment was located. This "public" demonstration of the invention resulted in the loss of NIH's patent protection for this invention.[5]

By their very nature, universities are public institutions. However, prudent universities and university research groups protect their IP by restricting who can see their inventions in practice. For example, research is conducted within the confines of the lab. Access to the lab is restricted to employees of the university or to persons who have acknowledged in writing appropriate non-disclosure policies of the university. Under these circumstances, the IP will be protected from loss of rights through an inadvertent public use bar.

FIRST TO INVENT

U.S. patent law is currently the only patent system that follows a first to invent regime.[6] Everywhere else in the world follows a first to file regime. Therefore, in every country in the world, except the U.S., the patent goes to the party that was the first to file its patent application irrespective of whether those individuals were the first to come up with the idea.

Under typical circumstances, there is very little difference between the systems as to who is entitled to the patent. Most typically, the first to invent also files first. However, notable exceptions to this general statement exist. For this reason, the current U.S. patent system has a special mechanism, the patent interference, to determine who is the first inventor.

Normally, the first to invent both conceived and reduced to practice the invention before the second inventor. However, interference law is far more complicated than that simple scenario. For example, it is not uncommon that the first to conceive was not the first to reduce to practice. Similarly, there are occasions where the first to conceive and the first to reduce to practice waited too long to file their

5. *Baxter Int'l, Inc. v. Cobe Labs, Inc.*, 88 F.3d 1054 (Fed. Cir. 1996)
6. The U.S. Congress recently passed legislation to change the U.S. patent system to a first-to-file regime. It is expected that by the time this chapter is published this change will be enacted into law.

patent application and then lost their patent rights, even though they were the first to arrive at the invention.

The key to success in patent interferences is good documentation. Without it, the first to invent could lose their rights to the patent. There are certain minimal requirements for what constitutes good documentation. Good documentation begins with reducing the inventive activities to writing. The writings should be dated and signed by the inventor. The writings should also be corroborated by a person who understands the invention, but is also not an inventor of the invention. The reason for the second person is that an inventor cannot self-corroborate the date of the invention.

Special consideration applies to electronic records. Courts understand that electronic dates can be manipulated. Therefore, in today's electronic society, companies need to have plans and methodologies for capturing electronic information in a way that will satisfy the courts' legal requirements. The most traditional method, and maybe most cost effective for smaller companies, is to simply print the electronic records and date and sign them as in any other paper record. This process, however, is not practical for many companies. Therefore, whether it is memorializing the electronic data onto a permanent media or in a relatively expensive electronic notebook system, the company should understand both what courts require and what is cost effective and available to that company.

INHERENCY

Another common problem health technology companies face regarding anticipation is the question of inherency. Under this doctrine, "[t]he discovery of a previously unappreciated property of a prior art composition, or of a scientific explanation for the prior art's functioning, does not render the old composition patentably new to the discoverer."[7] The inherent property need not be appreciated at the time of the invention. Under the law of inherency, a chemical composition and its properties are inseparable; therefore, if the prior art teaches the identical chemical structure, the properties disclosed and/or claimed are necessarily present.

This type of rejection is routinely made in the antibody field. For

7. *Atlas Powder Co. v. Ireco Inc.*, 190 F.3d 1342 (Fed. Cir. 1999)

example, where the invention pertains to a particular antibody having a particular therapeutic efficacy and generically disclosed in the prior art are antibodies to the same target, the USPTO typically takes the position that the earlier antibodies have the claimed therapeutic efficacy, absent a technical showing to the contrary. Obviously such rejections can be particularly difficult to address without engaging in expensive and, in the case where the antibody is not commercially available, potentially impossible experiments. The ability to address such rejections on the "paper" can be difficult, but it is not impossible. The key involves the skill of the advocate and the relative state of the science. In marshalling the facts about what the art taught about the prior antibodies, such anticipation rejections can be overcome without reverting to experimental evidence.

OBVIOUSNESS

To protect against trivial modifications of the prior art, the patent system also precludes granting a patent when the claimed invention would have been obvious to the person of ordinary skill in the art. 35 U.S.C. § 103 defines obviousness. The pertinent paragraph of Section 103 provides:

> (a) A patent may not be obtained though the invention is not identically disclosed or described as set forth in section 102 of this title, if the differences between the subject matter sought to be patented and the prior art are such that the subject matter as a whole would have been obvious at the time the invention was made to a person having ordinary skill in the art to which said subject matter pertains. Patentability shall not be negatived by the manner in which the invention is made.[8]

In comparing the claim to the prior art, three factual inquiries must be made: (1) the level of ordinary skill in the art at the time the invention was made must be evaluated; (2) the scope and content of the prior art at the time the invention was made must be ascertained; and (3) the differences between the claimed invention and the prior

8. 35 U.S.C. § 103(a)

art must be determined.

The person of ordinary skill in the art is a legal construction of a hypothetical person who presumptively has the full knowledge of those working in the field. The hypothetical person of ordinary skill in the art "would, of necessity, have the capability of understanding the scientific and engineering principles applicable to the pertinent art."[9]

In the second part of the inquiry, the prior art is compared to "the field of the inventor's endeavor" and "the particular problem with which the inventor was involved" at "the time the invention was made."[10]

The third inquiry, ascertaining the differences between the claim at issue and the prior art, considers the claimed invention and the prior art as a whole. This inquiry focuses on whether there is some rationale to combine or modify the references. The rationale can come from the references themselves or from the general knowledge of the person of skill in the art. The traditional test applied by the Federal Circuit was whether there was a teaching, suggestion, or motivation (TSM test) taught by the references. The Supreme Court recently clarified that the TSM test was too rigid. Thus, while the TSM test can still be applied, it is now no longer required that the references contain a teaching, suggestion, or motivation to combine. The Supreme Court listed the following principles as instructive when determining whether a patent claiming the combination of elements of prior art is obvious:

> When a work is available in one field of endeavor, design incentives and other market forces can prompt variations of it, either in the same field or a different one. If a person of ordinary skill can implement a predictable variation, § 103 likely bars its patentability. For the same reason, if a technique has been used to improve one device, and a person of ordinary skill in the art would recognize that it would improve similar devices in the same way, using the technique is obvious unless its

9. *Ex parte Hiyamizu*, 10 U.S.P.Q.2d 1393, 1394 (Bd. Pat. App. & Int. 1988)
10. *See Monarch Knitting Mach. Corp. v. Sulzer Morat GmbH*, 139 F.3d 877 (Fed. Cir. 1998)

> actual application is beyond his or her skill...[A] court must ask whether the improvement is more than the predictable use of prior art elements according to their established functions.[11]

The Supreme Court acknowledged the difficulty in applying these principles where "the claimed subject matter may involve more than the simple substitution of one known element for another or the mere application of a known technique to a piece of prior art ready for the improvement."[12] The Supreme Court stated:

> Often, it will be necessary for a court to look to interrelated teachings of multiple patents; the effects of demands known to the design community or present in the marketplace; and the background knowledge possessed by a person having ordinary skill in the art, all in order to determine whether there was an apparent reason to combine the known elements in the fashion claimed by the patent at issue. To facilitate review, this analysis should be made explicit. As our precedents make clear, however, the analysis need not seek out precise teachings directed to the specific subject matter of the challenged claim, for a court can take account of the inferences and creative steps that a person of ordinary skill in the art would employ.[13]

WHAT THE INVENTOR MUST DISCLOSE

The patent system is premised on an agreement between the government that in exchange for the public disclosure of how to make and use the invention the inventor receives the patent, which gives the inventor the right to exclude others from practicing the claimed invention. Important to this agreement is the obligation to disclose how to make and use the invention. The first paragraph of 35 U.S.C. § 112 defines the information an applicant must disclose.[14] This paragraph

11. *KSR v. Teleflex,* 550 U.S. 398 (2007) (internal citations omitted)
12. *Id.* at 417
13. Id.
14. 35 U.S.C. § 112 (2009)

contains three requirements and states:

> [t]he specification shall contain a written description of the invention, and of the manner and process of making and using it, in such full, clear, concise, and exact terms as to enable any person skilled in the art to which it pertains, or with which it is most nearly connected, to make and use the same, and shall set forth the best mode contemplated by the inventor of carrying out his invention.[15]

Thus, the three requirements are: (1) a written description requirement; (2) an enablement requirement; and (3) a best mode requirement. The consequence of failing to meet any of the three requirements of section 112, first paragraph, is that the claim at issue is invalid.

The Written Description Requirement

Under 35 U.S.C. § 112, first paragraph, the specification of a patent application must "contain a written description of the invention."[16] "The purpose of this provision is to ensure that the scope of the right to exclude, as set forth in the claims, does not overreach the scope of the inventor's contribution to the field of art as described in the patent specification."[17]

Generally, the written description requirement is met if the patent specification describes the invention "in sufficient detail that one skilled in the art can clearly conclude that 'the inventor invented the claimed invention.'"[18] The specification need not disclose the limitations of the claim in exactly the same words used in the claim. Rather, "[t]he test is whether the disclosure of the application relied upon reasonably conveys to a person skilled in the art that the inventor *had possession* of the claimed subject matter at the time of the earlier filing date."[19]

15. 35 U.S.C. § 112, first paragraph (2009)
16. 35 U.S.C. § 112 (2009)
17. *Reiffin v. Microsoft Corp.*, 214 F.3d 1342 (Fed. Cir. 2000)
18. *Regents of Univ. of Cal. v. Eli Lilly & Co.*, 119 F.3d 1559 (Fed. Cir. 1997)
19. *Eiselstein v. Frank*, 52 F.3d 1035 (Fed. Cir. 1995) (emphasis added)

To demonstrate possession, the application must "describe the claimed subject matter in terms that establish that the applicant was in possession of the...claimed invention, including all of the elements and limitations."[20]

However, "possession" is not the only requirement of written description. The specification also has to *describe* the invention.

For example, the disclosure of a species, or a limited number of species, does not necessarily provide an adequate written description of a broad genus which encompasses the species.[21] One way to satisfy the written description requirement is to disclose a "representative number of species within the genus."[22] A patent that does not disclose any compounds that can be used in its claimed methods lacks adequate written description.[23]

The Enablement Requirement

Under 35 U.S.C. § 112, first paragraph, the specification "shall contain a written description of the invention...as to enable any person skilled in the art to which it pertains, or with which it is most nearly connected, to make and use the same..."[24] The enablement requirement of section 112 is distinct from the written description requirement.

"The test for enablement is whether one reasonably skilled in the art could make or use the invention from the disclosures in the patent coupled with information known in the art without undue experimentation."[25] An application is enabling if one of ordinary skill in the art could have combined the teachings of the reference with his or her own knowledge to make and use the claimed invention.

To determine whether the claimed invention is enabled, the following factors are routinely considered: (1) the breadth of the claims; (2) the nature of the invention; (3) the state of the prior art; (4) the level of one of ordinary skill; (5) the level of predictability in the art;

---

20. *Univ. of Rochester v. G.D. Searle & Co.*, 358 F.3d 916 (Fed. Cir. 2004)
21. *University of Cal.*, 119 F.3d at 1568
22. *Id.* at 1569
23. *Univ. of Rochester*, 358 F.3d at 927
24. 35 U.S.C. § 112 (2009)
25. *United States v. Techtronics, Inc.*, 857 F.2d 778 (Fed. Cir. 1988)

(6) the amount of direction provided by the inventor; (7) the existence of working examples; and (8) the quantity of experimentation needed to make or use the invention based on the content of the disclosure.[26]

Disclosure of a species does not necessarily provide adequate enablement across an entire genus.[27] In *Genentech*, the patentees claim recited a method of producing a protein consisting essentially of amino acids 1-191 of human growth hormone to an accused composition consisting of amino acids 1-191 of human growth hormone, produced by a method comprising post-translational enzymatic cleavage of a fusion protein.[28] The specification, however, only suggested that cleavable fusion proteins were possible.[29]

> [W]hen there is no disclosure of any specific starting material or of any of the conditions under which a process can be carried out, undue experimentation is required; there is a failure to meet the enablement requirement that cannot be rectified by asserting that all the disclosure related to the process is within the skill of the art. It is the specification, not the knowledge of one skilled in the art, that must supply the novel aspects of an invention in order to constitute adequate enablement.[30]

Thus, novel aspects of an invention must be described in greater detail to satisfy the enablement requirement.

Biotechnology patents frequently face the issue of whether a specification that sets forth a single or a limited number of examples can enable broad claims when the subject matter concerns biological materials, which are generally considered to be unpredictable. For example, in *In re Vaeck*[31], the Federal Circuit upheld the PTO rejection of broad claims to chimeric genes capable of being

26. *In re Wands*, 858 F.2d 731 (Fed. Cir. 1988)
27. *Genentech, Inc. v. Novo Nordisk A/S*, 108 F.3d 1361 (Fed. Cir. 1997)
28. *Id.*
29. *Id.*
30. *Id.* at 1366.
31. *In re Vaeck*, 947 F.2d 488 (Fed. Cir. 1991)

expressed in cyanobacteria cells because the specification gave only two working examples using a single species. The Federal Circuit noted that there was a "relatively incomplete understanding of the biology of cyanobacteria as of appellants' filing date," and that "only one particular species of cyanobacteria is employed in the working examples."[32] Accordingly, the Federal Circuit held that "there is no reasonable correlation between the narrow disclosure in appellants' specification and the broad scope of protection sought in the claims encompassing gene expression in any and all cyanobacteria."[33]

Similarly, in *Enzo Biochem, Inc. v. Calgene, Inc.*[34], the Federal Circuit applied the *Wands* factors to the use of antisense technology in both eukaryotic and prokaryotic cells. Although the specification indicated that "[t]he practices of this invention are generally applicable with respect to any organism containing genetic material which is capable of being expressed," the only working examples provided were for three genes in the prokaryote *E. coli.*[35] The Federal Circuit concluded that "[Enzo] has provided only the mere 'germ of the idea' for exploiting antisense in eukaryotes... What is glaringly 'missing' from the specifications is the disclosure of any direction or examples of how such an idea might be implemented in any other cell than *E. coli.*"[36]

In concluding that the claims were not enabled over their full scope, the Federal Circuit relied on evidence of multiple failed attempts by the inventor to achieve success, even in *E. coli.* Furthermore, during prosecution of the patent, Enzo had submitted a declaration and post-filing art in support of enablement to demonstrate that practicing the invention in eukaryotic systems was "routine." Noting that these references were published in highly esteemed journals such as *Nature* and *Cell*, the Federal Circuit pointed out that it was highly unlikely that these results were "routine." Accordingly, the Federal Circuit concluded that these articles, even though demonstrating successful practice of antisense technology in eukaryotic cells, were "as much a suggestion of nonenablement as enablement."

32. *Id.*
33. *Id.*
34. *Enzo Biochem, Inc. v. Calgene, Inc.* 188 F.3d 1362 (Fed. Cir. 1999)
35. *Id.*
36. *Id.*

The Federal Circuit has expressed difficulty "allowing an inventor to claim what was specifically desired but difficult to obtain at the time the application was filed, unless the patent discloses how to make and use it."[37] In *Plant Genetic Systems*, the key issue was whether the claims covered monocotyledons (monocots) and dicotyledons (dicots), when the specification only disclosed working examples of dicots. The Federal Circuit affirmed the district court's conclusion of non-enablement, noting that

> monocots and stably transformed monocot cells were not an unknown concept that came into existence only after [the time the application was filed]. But stably transformed monocot cells were difficult to produce, and the [] patent gave no instruction how.[38]

Another informative case involving the filed of biotechnology is *Chiron Corp. v. Genentech, Inc.*[39] In *Chiron*, the claims related to a monoclonal antibody that binds to c-erbB-2 (HER2), a breast cancer antigen.[40] The claims, however, covered murine and chimeric antibodies. The Federal Circuit examined two of Chiron's applications, filed in 1985 and 1986, and concluded that the claims were not supported by those applications because the specifications only disclosed murine monoclonal HER2 antibodies and contained no guidance regarding chimeric HER2 antibodies. The Federal Circuit observed that "[w]hile Chiron's applications certainly enable murine antibodies, they do not enable chimeric antibodies."[41]

Although it seems from the above that biotechnology patents routinely suffer from enablement problems, not all do. In *Amgen v. Hoechst*, 314 F.3d 1313 (Fed. Cir. 2003), the Federal Circuit affirmed that claims directed towards vertebrate cells which "'can be propagated *in vitro*,' comprise 'non-human DNA sequences that control transcription,' transcribe 'DNA encoding human erythropoietin,' and produce the claimed amount of EPO" were enabled based on a

---

37. *Plant Genetic Sys., N.V. v. DeKalb Genetics Corp.*, 315 F.3d 1335 (Fed. Cir. 2003)
38. *Id.*
39. *Chiron Corp. v. Genentech, Inc.*, 363 F.3d 1247 (2004)
40. *Id.*
41. *Id.*

disclosure of two different cell lines which met the limitations.[42] The district court examined, among other evidence, expert testimony that EPO could be easily expressed in any vertebrate cell.[43]

In summary, the enablement requirement dictates that a patent specification disclose sufficient information so that one of skill in the art can practice the claim over its full breadth without undue experimentation.

THE BEST MODE REQUIREMENT

Section 112, first paragraph, provides that "[t]he specification...shall set forth the *best mode* contemplated by the inventor of carrying out his invention."[44] The Federal Circuit has explained that the purpose of the best mode requirement is:

> [T]o ensure that a patent applicant plays "fair and square" with the patent system. It is a requirement that the *quid pro quo* of the patent grant be satisfied. One must not receive the right to exclude others unless at the time of filing he has provided an adequate disclosure of the best mode known to him of carrying out his invention.[45]

The *quid pro quo* of the patent system is the government's limited grant of exclusive rights in the invention in exchange for the inventor's full disclosure of the invention to the public. Thus, an applicant may not apply for a patent while concealing a mode known by the applicant to be better than that disclosed.[46]

Simply put, the best mode requirement of section 112, first paragraph, imposes an obligation upon a patent applicant that cannot be satisfied by merely disclosing the invention in a patent application. What is required is the disclosure of preferred methods for carrying out the invention. Thus, "[a] patent applicant must disclose the *best* mode of carrying out his claimed invention, not [simply] *a* mode of

42. *Id.*
43. *Id.*
44. 35 U.S.C. § 112, first paragraph (2009) (emphasis added).
45. *Amgen, Inc. v. Chugai Pharm. Co., Ltd.*, 927 F.2d 1200 (Fed. Cir. 1991)
46. *Spectra-Physics, Inc. v. Coherent, Inc.*, 827 F.2d 1524 (Fed. Cir. 1991)

making and using what is claimed."[47]

The best mode requirement is a subjective standard that requires disclosure of the best mode as contemplated by the inventor, not the absolute (*i.e.*, objective) "best" mode.

Determination of a patentee's satisfaction of the best mode requirement involves a two-pronged inquiry. In *Chemcast Corporation v. Arco Industries Corporation*,[48] the Federal Circuit set forth a two-part test for determining whether a best mode violation has occurred. The first inquiry is whether the inventor, at the time of filing the application, contemplated a best mode of practicing the claimed invention. This inquiry is a subjective one. If the first inquiry yields an affirmative answer, the second (objective) inquiry is whether the disclosure is adequate to enable one of ordinary skill in the art to practice the best mode or, rather, whether the best mode has been concealed from the public. If concealment of the best mode is found, all claims of the patent are rendered invalid.[49]

TRADE SECRETS

Trade secrets are generally defined by state law, with most states having adopted the Uniform Trade Secrets Act (UTSA). Generally speaking, a trade secret is any type of knowledge, information, or device used in one's business to confer a competitive advantage over competitors who do not know or use it. Under the UTSA, a trade secret includes information, including a formula, pattern, compilation, program device, method, technique, or process, that: derives independent economic value, actual or potential, from not being generally known to, and not being readily ascertainable by proper means by, other persons who can obtain economic value from its disclosure or use, and is the subject of efforts that are reasonable under the circumstances to maintain its secrecy.

Although patents have many advantages, trade secret protection has certain elements that are not available to patents. For example,

47. *Chemcast Corp. v. Arco Indus. Corp.*, 913 F.2d 923 (Fed. Cir. 1990) (emphasis in original)
48. *Chemcast Corporation v. Arco Industries Corporation*, 913 F.2d 923 (Fed. Cir. 1990)
49. According to the recently passed legislation changing the U.S. patent system, best mode violations will no longer result in a finding of invalidity or unenforceability by a court of law.

while patents generally last only twenty years, trade secrets can last forever. Think, for example, of the secret formula of the Coca-Cola syrup. Unlike patents, trade secrets require no filing or maintenance costs. Also, whereas patents may take some time to issue, trade secrets are enforceable the moment they are created.

This is not to imply that trade secrets have no disadvantages, as they have several discrete disadvantages. Depending on the nature of the secret, it may be possible to reverse engineer and discover the secret. Especially as technology is ever evolving, the trade secret that was previously safe from reverse engineering may become vulnerable as new analytical techniques become discovered. Thus, once the secret is public, the trade secret protection is voided.

For companies considering utilization of trade secrets as an IP tool, several best practices are recommended. The company should set up measures to protect against disclosure. Employment agreements should discuss the importance of trade secrets and the employee's obligations not to disclose the same. Key trade secrets should be identified and safeguards should be established to prevent their disclosure. The company should consider restricting access to facilities housing trade secrets. Similarly, trade secrets should be disclosed on a need-to-know basis and if possible information about the trade secret should be compartmentalized across different employees. The company should have a training program that builds an atmosphere of respect for trade secrets and also reminds the employees of the importance of trade secrets to the company's success.

Although patents provide great value to health technology companies, trade secrets may provide even more value under the appropriate circumstances. Specifically, whether a product can or cannot be easily reverse engineered may be the most important consideration in evaluating the potential strength of a trade secret. If the product cannot be easily reverse engineered, then trade secret protection may be best. In the context of biotechnology-based inventions, there are many steps used in the manufacture and purification of large complex biological molecules, such as antibodies, that cannot be easily reverse engineered. Therefore, whereas patenting the bio-molecule may make most sense, it may make equal sense to hold process information as a trade secret.

CONCLUSION

Because of the importance of intellectual property to the overall success of companies in the health technology industry, proper care must be taken in both the procurement of intellectual property and dealing with the intellectual property of others.

Chapter 6

Small Business Intellectual Property Considerations When Doing Business With the Federal Government

William J. Bainbridge

William Bainbridge is an attorney at Cohen Mohr LLP, in Washington, D.C. His practice focuses on all aspects of government contracts law, including bid protests, claims and disputes. Mr. Bainbridge has experience litigating bid protests before the Government Accountability Office (GAO) and the Court of Federal Claims. Prior to his legal career, Mr. Bainbridge served in the U.S. Army for over 22 years. Mr. Bainbridge can be contacted at bbainbridge@cohenmohr.com.

In many ways, there has never been a better time for high tech small businesses looking to expand into the federal marketplace. Two programs in particular represent the kinds of opportunities available for small business entrepreneurs, namely the Small Business Administration's ("SBA") Small Business Innovation Research ("SBIR") and the Small Business Technology Transfer ("STTR") programs. The government relies on the SBIR and STTR programs to promote opportunities for high tech small businesses to develop their technologies under contract with various federal agencies.[1] Since 2004, when President George W. Bush signed an Executive Order[2] aimed at reinvigorating the government's commitment to the SBIR and STTR programs, the number of programs and the amount

1. United States Small Business Administration, *About SBIR and STTR Programs*, available at www.sbir.gov/about/index.htm
2. Executive Order 13329, 69 Fed. Reg. 38, 1981-2 (Feb. 24, 2004).

of money agencies spend on these contracts has only increased.[3] And just this year, the government amended the SBIR program, raising the SBIR Phase I and Phase II award threshold amounts from $100,000 to $150,000, and $750,000 to $1,000,000 respectively.[4] For those unfamiliar with the SBIR and STTR programs a brief description is in order.

OPPORTUNITIES FOR SMALL BUSINESSES

Under the SBIR program, which is broken down into three phases, the government solicits proposals from small technology businesses for early stage research and development ("R&D") work.[5] Typically, Phase I work focuses on determining "the scientific, technical, and commercial merit and feasibility of ideas submitted under the SBIR Program. In other words, during Phase I the government is looking to determine whether the potential for further development exists. Phase I award amounts are around $150,000, as indicated above, and the contracts generally are for six to nine months. Current solicitations may be found at http://www.acq.osd.mil/osbp/sbir/solicitations/.

Phase II SBIR awards are based on the results achieved during Phase I, as well as the merit of the firm's Phase II proposal for continued development. Award amounts are around $1,000,000, as indicated above, and the contract period runs for about 24 months. Successful Phase II awardees may ultimately move into Phase III, under which "the small business is expected to obtain funding from

3. For example, the Department of Defense ("DoD") made SBIR/SSTR awards valued at approximately $1,222,858,799 in 2009, up from $894,949,718 in 2003. DOD, SBIR Annual Report, available at www.dodsbir.net/annualreport/annrpt.html. Similarly, the SBIR/STTR awards at National Institute of Health ("NIH") went up from $562,423,596 to $651,344,415. NIH, *SBIR/STTR Award Data*, available at http://grants.nih.gov/grants/funding/award_data.htm.
4. 75 Fed. Reg. 60, 15756-7 (March 10, 2010).
5. For simplicity and consistency, this article will focus small business considerations with the Department of Defense, which has the largest share of the federal government's small business contracts. References to small business programs, in general, will cite to DoD programs and regulations. References to other federal agencies will be made expressly as required. It is important to note that agencies often have very different goals for their grant programs. For example, the National Science Foundation awards grants to high impact, high risk research without having topics of interest or a motivation to share technologies with competitors.

the private sector and/or non-SBIR Government sources to develop the prototype into a viable product or non-R&D service for sale in military and/or private sector markets."

Each year, government agencies announce their topics of interest for the SBIR program. For example, the Army's current topic list includes:

- Detection of Contaminants in Aviation Fuel
- Near Infrared Stretched Pulse Processing
- Deterministic Finishing of Domes with Buried Electromagnetic Structures
- High Density Liquid Hydrogen Missile
- Fuel Detection and Serotype Identification of Dengue Virus in the Mosquito Vector
- Software Tool for Complex Biomarker Discovery
- Controlled Release of Topical Nitric Oxide for Treating Cutaneous Injuries
- Electromagnetic Interference Shielding Fabrics for Use with Soft Walled Shelters
- High Barrier Packaging Based on Melt Extrudable Liquid Crystal Polymers
- Fabric with Variable Air Permeability for Use in Parachutes
- Energy Efficient Ice Supply in Theatre (EEISIT)
- Improved Ballistic Combat Hearing Protection

At the same time, the Defense Advanced Research Projects Agency ("DARPA") included the following in its 2010 SBIR topic list:

- Low Noise, High Efficiency Hydraulics for Mobile Robots
- Robot Fabrication via Layered Manufacturing
- Global Intelligence, Surveillance, and Reconnaissance (ISR)
- Lithography-free Manufacturing of Polymer Photonic

Devices
- Wavelength-Stabilized, High-Brightness Diode Laser Pumps for High-Power Fiber Lasers
- Novel Acoustic Materials for Passive Hearing Protection
- Waveform Design, Database, and Development for Radio Communications
- High Speed Naval Surface Munitions

The point is this: our government is interested in—and is willing to fund—several different kinds of research into a broad array of scientific and engineering disciplines. Firms interested in obtaining one of these opportunities should begin by contacting the SBA and small business advocates at the various federal agencies that participate in the SBIR and SSTR programs.[6] Many small businesses also find that it is beneficial to contact an attorney who understands and can navigate the myriad contractual and regulatory requirements that are inherent in every government contract.

UNDERSTANDING YOUR CONTRACT

Any small business looking to develop a new technology under a contract with the government needs to know how to safeguard its intellectual property. In order to do this, prospective contractors must understand how *the contract* allocates the Government's rights in intellectual property. For example, your contract should contain language addressing rights in "technical data."[7] Technical data encompasses virtually any "recorded information, regardless of the form or method of the recording, of a scientific or technical nature

6. A list of federal agencies (with links) participating in the SBIR and SSTR programs may be found at the SBIR.gov website. Available at www.sbir.gov/federal_links.htm.
7. Rights in technical data are not the only kind of intellectual property concerns that could be addressed in your contract. If applicable, the contract may also specify your ability to obtain patents, copyrights, and trademarks. Depending on the nature of the contract, these may be negotiable. That said, my discussion will focus on the allocation of rights in technical data because it is relevant to large number of government contracts, and as such, provides a framework for a discussion on how Government contracts treat intellectual property in general.

(including computer software documentation)."[8] Equally important, however, the term "does not include computer software or data incidental to contract administration, such as financial and/or management information."

Rights in technical data are generally allocated according to which party funded their development.[9] Accordingly, the Government obtains one of three types of licenses, or usage rights, for its contractors' technical data. These are: "unlimited rights"; "Government purpose rights"; and "limited rights. ".[10] *Unlimited rights* permit the Government "to use, modify, reproduce, perform, display, release, or disclose technical data in whole or in part, in any manner, and for any purpose whatsoever, and to have or authorize others to do so." *Government purpose rights* include the right to... [u]se, modify, reproduce, release, perform, display, or disclose technical data within the Government without restriction; and... [r]elease or disclose technical data outside the Government and authorize persons to whom release or disclosure has been made to use, modify, reproduce, release, perform, display, or disclose that data for U.S. government purposes." *Limited rights* entail the "rights to use, modify, reproduce, release, perform, display, or disclose technical data, in whole or in part, within the Government." That said, the Government is not supposed to, without written permission by the contractor asserting limited rights, "release or disclose technical data outside the Government, use the technical data for manufacture, or

8. 48 C.F.R. § 252.227-7013. Similar language appears at 48 C.F.R. § .227-7018, which pertains to rights in non-commercial technical data and computer software in a SBIR contract. It is important to note that § 48 C.F.R. § 252.227-7013 covers the Government's rights in technical data for noncommercial items; there are other types of items and documentation for which the Government claims certain usage rights. In the interest of brevity, this article will only address rights in technical data for noncommercial items, which is illustrative of the usage rights addressed in most Government contracts.
9. *Lockheed Martin Corp. v. United States* , 42 Fed. Cl. 485, 495 n.7 (1998) ("...the source of funding for the research determines whether the government will obtain limited or unlimited rights in technical data generated by the research."), *aff'd in part, reversed in part on other grounds,* 210 F.3d 1366 (Fed. Cir. 2000).
10. 48 C.F.R. § 252.227-7013. Again, this article will focus on the DoD rules. Civilian agencies adhere to the Federal Acquisition Regulations, which provides that contractors are only required to supply form, fit and function data. If a civilian agency requires the contractor to provide limited or restricted rights data under 48 C.F.R. § 52.227-14, then the contract should include instructions for marking the data.

authorize the technical data to be used by another party..." At the same time, the Government has carved out exceptions to this rule for emergencies and various other Government interests.

According to the Defense Department's procurement regulations, the Government obtains unlimited rights in technical data that are:

(i) ...pertaining to an item, component, or process which has been or will be developed exclusively with Government funds;
(ii) Studies, analyses, test data, or similar data produced for this contract, when the study, analysis, test, or similar work was specified as an element of performance;
(iii) Created exclusively with Government funds in the performance of a contract that does not require the development, manufacture, construction, or production of items, components, or processes;
(iv) Form, fit, and function data;[11]
(v) Necessary for installation, operation, maintenance, or training purposes (other than detailed manufacturing or process data);
(vi) Corrections or changes to technical data furnished to the Contractor by the Government;
(vii) Otherwise publicly available or have been released or disclosed by the Contractor or subcontractor without restrictions on further use, release or disclosure, other than a release or disclosure resulting from the sale, transfer, or other assignment of interest in the technical data to another party or the sale or transfer of some or all of a business entity or its assets to another party;
(viii) Data in which the Government has obtained unlim-

11. *See* 48 C.F.R. § 27.401 ("'Form, fit, and function data'" means data relating to items, components, or processes that are sufficient to enable physical and functional interchangeability, and data identifying source, size, configuration, mating and attachment characteristics, functional characteristics, and performance requirements. For computer software it means data identifying source, functional characteristics, and performance requirements, but specifically excludes the source code, algorithms, processes, formulas, and flow charts of the software.")

ited rights under another Government contract or as a result of negotiations; or

(ix) Data furnished to the Government, under this or any other Government contract or subcontract thereunder, with-
   (a) Government purpose license rights or limited rights and the restrictive condition(s) has/have expired; or
   (b) Government purpose rights and the Contractor's exclusive right to use such data for commercial purposes has expired.

The important thing to take away from this language is this: if what you are developing for or providing to the Government was paid for *exclusively* with Government funds, the Government may assert unlimited rights to the technical data. In such cases, the Government will have the absolute right "to use, modify, reproduce, perform, display, release, or disclose technical data in whole or in part, in any manner, and for any purpose whatsoever, and to have or authorize others to do so."[12]

When the Government is not able to assert unlimited rights in technical data, it will want to assert Government purpose rights. Government purpose rights last for five years, unless you negotiate something different in your contract. Government purpose rights exist for technical data pertaining to:

(A) …items, components, or processes developed with mixed funding except when the Government is entitled to unlimited rights in such data…; or

(B) Created with mixed funding in the performance of a contract that does not require the development, manufacture, construction, or production of items,

12. 48 C.F.R. § 252-227-701. The Government may also obtain unlimited rights in the technical data submitted with a successful proposal. *See* 48 C.F.R. § 52.227-23 and 48 C.F.R. § 252.227-7017. At DoD, offerors must identify in their proposals all technical data and any restrictions (e.g. limited rights, restricted rights) in accordance with specific instructions. Failing to adhere to these rules when submitting a proposal may prevent a contractor from asserting its rights once the contract has been awarded. These rules do not apply outside of DoD.

components, or processes.

(i) ...Upon expiration of the five-year or other negotiated period, the Government shall have unlimited rights in the technical data.

(ii) The Government shall not release or disclose technical data in which it has government purpose rights unless—

(a) Prior to release or disclosure, the intended recipient is subject to the non-disclosure agreement at 227.7103-7 of the Defense Federal Acquisition Regulation Supplement (DFARS); or

(b) The recipient is a Government contractor receiving access to the data for performance of a Government contract that contains the clause at DFARS 252.227-7025, Limitations on the Use or Disclosure of Government-Furnished Information Marked with Restrictive Legends.

(iv) The Contractor has the exclusive right, including the right to license others, to use technical data in which the Government has obtained government purpose rights under this contract for any commercial purpose during the time period specified in the government purpose rights legend prescribed in paragraph (f)(2) of this clause.

Thus, if what you are producing for the Government is paid for partially with Government funds, then the Government may assert Government purpose rights. But beware. The Government retains the right to release your technical data to third parties, including other contractors (including your competitors). While Government purpose rights may provide some safeguards to your technical data, it is important to remember that when this clause is invoked, it is done so further the Government's interests (which may include the performance of a contract to which you are not a party), and not necessarily to protect any proprietary interests. And to the extent the Government has claimed Government purpose rights, those rights will transform into unlimited rights at the time specified in your contract.

Finally, there are limited rights. Limited rights arise when the data were developed exclusively with private funds. Moreover, the contractor desiring to assert limited rights must have taken steps to protect or maintain the data as proprietary information (or trade secret). And finally, the contractor asserting limited rights must mark the data with the "limited rights legend." In other words, before the Government will treat the data as proprietary, you must demonstrate that you have protected it as proprietary. But more on that later.

When the Government has limited rights in technical data, it has "the rights to use, modify, reproduce, release, perform, display, or disclose technical data, in whole or in part, within the Government." Accordingly, the Government is prohibited from releasing or disclosing the technical data without the written consent of the contractor asserting those rights. However, the Government has once again provided itself exceptions to the rule, namely, it may release technical data to third parties for emergency repairs and maintenance; or if the release technical data is required by a foreign government for evaluational or informational purposes. Unfortunately, neither one of these situations is well-defined, so in deciding whether one of these circumstances apply, the Government may exercise the broad discretion it has reserved to itself.

PROTECTING TECHNICAL DATA

As indicated above, in order to limit the Government's rights in technical data, contractors must mark the data according to the regulations. This issue was litigated in the U.S. Court of Federal Claims in *Night Vision Corp. v. United States.*[13] In that case, Night Vision Corporation was a recipient of a SBIR contract under which it had delivered prototype night vision devices to the U.S. Air Force. After accepting the prototype devices, the Air Force decided not to award Night Vision a follow-on contract for production. Instead, the Air Force decided to conduct a competitive procurement and provided a pair of Night Vision's goggles to competitors, allowing those

13. *Night Vision Corp. v. United States* , 68 Fed. Cl. 368 (2005).

firms to inspect the prototypes for three hours.[14] While Night Vision properly marked all of its data in accordance with the SBIR data rights legend, it did not mark the actual goggles and the box which contained them.[15] With respect to Night Vision's ability to limit the Government's rights in the goggles, the Court held:

> By its own language, § 252.227-7018(f) indicates that a contractor may only restrict the government's use and disclosure of technical data by marking the deliverable data with the appropriate data legend. A failure to use the appropriate legend results in the government receiving complete, unrestricted use. Other sections of the FAR which limit the government's rights in proprietary data developed by contractors have consistently been interpreted in this vein. *See Ervin & Assocs., Inc. v. United States,* 59 Fed.Cl. 267 (2004) (notwithstanding contractor's "oral statements, letters, and e-mails" allegedly asserting protection of proprietary data, the contractor's failure to mark data with the "Limited Rights Notice" prescribed in 48 C.F.R. § 52.227-14 resulted in government obtaining unlimited data rights); *Gen. Atronics Corp.,* ASBCA No. 49196, 2002 WL 450441 (Mar. 19, 2002); *E.M. Scott & Assocs.,* ASBCA No. 45,869, 94-3 BCA ¶ 27,059, 1994 WL 409331 (July 28, 1994) (contractor's failure to use the restrictive legend of 48 C.F.R. § 52.215-12 gave government unrestricted rights to data); *Bell Helicopter Textron,* ASBCA No. 21,192, 85-3 BCA ¶ 18,415, 1985 WL 17050 (Sept. 23, 1985) (contractor's failure to mark technical engineering drawings with a restrictive legend required by a contract clause similar to 48 C.F.R. § 252.227-7013 resulted in the government receiving unlimited rights to the data conveyed by the

14. *Id.* at 371-377. The DFARS data rights clause at issue in *Night Vision* pertained to SBIR contracts; the marking requirements, however, are similar to DoD's data rights clauses applicable to non-SBIR contracts. *See* 48 C.F.R. § 252.227-7013 and 48 C.F.R. § 252.227-7014.
15. *Id* . In denying Night Vision's claims, the Court apparently equated the goggles with "data."

drawings).[16]

Based on this framework, the Court concluded:

> [I]t is undisputed that neither the prototypes that plaintiff delivered to defendant nor the packaging in which those prototypes arrived were marked with proprietary data legends... This failure to use an appropriate data legend gave the government unrestricted use to the... prototypes. Plaintiff's argument that it was sufficient to have affixed the legend to the "technical drawings and documentation" associated with the goggles is not persuasive. To allow an exception to § 252.227-7018 as the plaintiff advocates would be contrary to the language of the regulation, contradict its intent, and ultimately render the regulatory protection of data unworkable. ***There simply is no provision in 48 C.F.R. § 252.227-7018 for the extension of the data rights protection from properly marked data to unmarked delivered data...***
>
> Restrictive legends alert all government officials-even those unfamiliar with the data rights of the contractor-that data is considered proprietary and is inappropriate for dissemination. Creating exceptions to the restrictive rights requirement would place government officials in the difficult position of being unsure which data was subject to restrictions and which was not...
>
> Since plaintiff did not attach the SBIR data rights legend to the...prototypes, no restrictions applied to defendant's use of the prototypes. [The Government] was thus within its rights when it shipped the...prototypes

16. 68 Fed.Cl. at 380.

to [the competitor]...[17]

In *Night Vision*, the contractor incorrectly thought that its rights in technical data were protected under its SBIR contract. However, when the Air Force elected not to award a SBIR Phase III contract, the Government was free to provide the prototypes to Night Vision's competitors in order to solicit competitive proposals—even though Night Vision had done the work of developing much of the technology.

Night Vision is a cautionary tale for small businesses. To protect your interest in technical data, remember the following:

- *Read your contract* and understand how rights in technical data are allocated;

17. *Id*. at 380-381 (emphasis added). The Court dropped included an interesting footnote questioning whether Night Vision's prototype was entitled to any protection at all:

> While not deciding this issue, it is likely that even if plaintiff had affixed data rights legends to the goggles, or if the notice provided for the technical drawings and documentation could extend to related items like the goggles, defendant would still be entitled to summary judgment because it is doubtful that the goggles constituted "data" eligible for protection under 48 C.F.R. § 252.227-7018(b)(4)(ii). FAR § 52.227-20(a) defines "Data" as used in § 252-2277018 as "recorded information, regardless of form or the media on which it may be recorded." 48 C.F.R. § 52.227-20(a)...
> Thus, a tangible product delivered under a contract may indeed constitute the embodiment of data, but it is not itself recorded information. Indeed, expert commentators have opined that "***the Government may legitimately provide a sample of a product to another company with full knowledge that it will be 'reverse engineered' to learn how to make a duplicate, even if the government may not provide the technical data associated with the object***." ...("[U]nless the item is covered by a patent, or is otherwise restricted by contract, the Government may use the item to reverse engineer its design or may provide it to a competitor to do the same.")...
> *Id*. at 381 n.16 (internal citations omitted) (emphasis added).

According to the Court, reverse engineering probably would have been permissible; however, reverse engineering would not be permissible if the product is protected by patent. *See* 10 U.S.C. § 2302(a)(1) and 41 U.S.C. § 418a(a) (prohibiting the Government and its contractors from interfering with patents or copyrights or any other right in technical data established by law). Moreover, the Government's right to reverse engineer is limited to obtaining spare parts when technical data is not available. 48 C.F.R. § 217.7504.

- Contact your attorney early in the process;
- Data rights are generally allocated according to which party paid for their development;
- If you develop technical data exclusively at private expense, you must properly treat the data as proprietary (e.g. markings in accordance with the Government's requirements) before you can assert limited rights on the Government;
- The Government has carved out exceptions for itself. The Government's interests will (almost) always trump those of its contractors;
- There is nothing prohibiting contractors from negotiating language in their contracts to protect proprietary rights.

Remember, although the federal government provides excellent opportunities for small businesses to develop new technologies, contracting with the government can be fraught with pitfalls. You can avoid many of these pitfalls by following these simple recommendations and keeping your government contracts attorney involved from the beginning.

Chapter 7

Standard Setting Organizations

Marlin Blizinsky

Marlin Blizinsky previously worked in the Corporate Standards and Interoperability Group within the Microsoft Legal and Corporate Affairs Department. There he advised product groups and worked on corporate intellectual property policy and standards issues including IP issues in China. Since leaving Microsoft, Mr. Blizinsky has expanded his focus to include federal and regional communications policy. Mr. Blizinsky lives in Seattle and enjoys the technology ferment in the area. Mr. Blizinsky can be contacted at marlinblizinsky@hotmail.com.

In the late fall of 1979, my wife and I were awakened by the following message over the loud speaker as we slept on the train from Paris to Lisbon: "Please don't step out of your car." The train tracks in France were a different gage than those in Spain. For the next six hours the railroad company lifted up each car one-by-one and replaced the wheels usable in France with those usable in Spain and Portugal.

Wikipedia lists ten different meanings for the word "standard." This chapter uses the term to mean a written specifications for computer interfaces that are language and platform neutral and are designed on a consensus basis by a group to enable information to pass between two or more pieces of equipment or software. Groups that write standards use a variety of terms for their work product including recommendations, specifications, and Request for Comments.

Some observers use the term *de facto* standard to mean "a custom, convention, company product, corporate standard, etc. which becomes generally accepted and dominant…" This article does not discuss *de facto* standards. Nor does it discuss process quality standards, such as ISO 9000. Finally, it does not discuss those additional issues associated with standards that are adopted as regulations.

STANDARDS

In the computer and communications world, most standards are interfaces. As such, they are designed to be platform and language neutral. The U.S. government defines an interface standard as: "A standard that describes one or more functional characteristics (such as code conversion, line assignments, or protocol compliance) or physical characteristics (such as electrical, mechanical, or optical characteristics) necessary to allow the exchange of information between two or more (usually different) systems or equipment." Standards are intended to make something more usable by making the hardware, software, or some combination easier to use or work together.

Some industries produce multiple, competing standards. One of the most interesting modern standards battles was between the tape formats used for home playback of movies on magnetic tape, Betamax and VHS. When movies were offered in these competing formats sales were slow, prices were high. Potential customers were reluctant to spend a significant amount of money buying movies fearing that the format they purchased would disappear. VHS eventually triumphed; studios stopped releasing movies in Betamax. Sales of VHS players soared, the cost of buying players and movies dropped, and the market for movies exploded.

Some markets are able to sustain more than a single standard. The world currently has two major cell phone technologies, GSM and CDMA. In addition, Japan has its own cell phone technology used nowhere else. These standards will soon be replaced by other fourth generation (4G) technologies including Wi-Max and Long Term Evolution or LTE.

There are a great many well-known standards including Blu-ray Disc, Wi-Fi, USB 2.0, MP3, GSM, and HTML to name just a few. Standards organizations around the world produce hundreds of standards a year, many of which are known only to specialists in the field. For example, the Advanced Access Content System (AACS) is a standard for copy protection and distribution of material stored on Blu-ray Discs. Despite the widespread familiarity with the term Blu-ray and despite this association, AACS is not mentioned in commercials for new movie disk releases and would be an unknown term even for most owners of Blu-ray Disc players.

Standards development, the writing of a new or modified specification, is one form of joint technical development. There are others including partnerships and joint ventures. Many of the challenges and legal issues are similar.

The development of a standard is related with to, but not the same as, the inclusion of the standardized technology in a product or component. Importantly, a company usually does not need to have participated in the development of a standard to implement that standard. Equally, participation in developing a standard does not legally obligate a participant to implement the standard in its products.

Standards vary in their level of specificity, i.e. some standards are written so there is a great deal of room for interpretation in implementation; others are written with little latitude. If there is room for interpretation, developers may interpret them differently and two implementations may to both comply with a standard but not work together. If there is no latitude, all implementations are the same and there is no differentiation. One common standard where this occurs interoperability has been a problem is Voice over Internet Protocol (VoIP). In situations where products must work together, you must not only test that the implementation complies with the standard, you must also have the products tested for interoperability. When interoperability is important, testing facilities will provide such service or the standards body will hold events, often called a plugfest, where vendors can bring their products and connect them to other products that also implement the standard.

Some authors use the term "standard" to describe an implementation, and some groups that call themselves standards development organizations produce implementations. As used in this article, a standard is a set of requirements; the implementer is free to decide how best to meet those requirements in his or her product. As Wikipedia defines it, a standard "describes *what* the system should do, not (necessarily) *how* the system should do it."

An implementation is one example of how to meet the requirements of the standard. In a competitive market, products are differentiated by the quality of their implementations as well as by their price, time to market, or other features.

REASONS FOR PARTICIPATING IN STANDARDS DEVELOPMENT ORGANIZATIONS

The decision whether to participate in an SDO is just another business decision. The following are some of the reasons participating in standards development may be worthwhile:

- To gain knowledge to assist your implementation
- To encourage the development of a standard that is compatible or not incompatible with your product
- To get a patent into the standard
- To fulfill the request of a partner
- To meet potential partners
- To benefit from or participate in standards-related marketing activities

Participating in the development of a standard can give participants a deep understanding of the standard. This can be a major benefit if you plan to implement a standard, especially one that is outside your normal competence or is unusually complex. Membership does not include an obligation to implement a standard, however.

Participating in the development of a standard gives participants information about the standard before nonparticipants have it. This may enable you to begin implementation before the standard is complete and before nonmembers have sufficient information to begin their work, thus enabling members to get compliant products to market earlier than competitors. Some companies appear to make use of this strategy in many of their products. There is obviously a degree of risk in implementing a specification that is not yet final, however.

It may be worthwhile to consider participating in standards development work with the goal of obtaining a standard that is compatible with, or at least not incompatible with, your product. This is especially true if you must implement the standard because of market requirements. While standards are platform and language neutral in theory, some specifications are easier to implement in some environments than in others. A simple example is the amount of processing power needed to implement the standard: cell phones typically have less computing power than computers.

It may be beneficial to have an essential patent in the standard. If your company will be implementing the standard, you will have something of value to trade with other essential patent holders who also want to implement. This also allows you to participate in patent pool discussions. In some instances, licensing your patent for standards implementation may be a significant source of revenue. However, even if the standard will be implemented without patent licensing fees, having a patent claim in the standard gives you leverage in negotiations over the other license terms such as the right to a reciprocal license on the same terms from implementers.

A partner may request that you participate in a Standards Development Organization (SDO) to support its efforts. This request may be limited to voting or the request may be more substantial. A partner may also ask that you support its membership in an SDO where membership is not open. Before accepting such an invitation, you will want to determine whether your interests in the organization are compatible and what you will do if they are not.

SDO participation can bring you into contact with potential partners. While some SDO activities will be made up solely of competitors, major development activities will often attract participants from multiple segments of the product ecosystem. The positions participants take during standards development may suggest opportunities for implementation, marketing, or other joint activity outside the organization.

Lastly, your participation in an important standards activity may add credibility to your product or company.

REASONS FOR NOT PARTICIPATING IN STANDARDS DEVELOPMENT ORGANIZATIONS

Participating in standards development costs resources. Thus, the choice whether to participate in standards development is just another cost/benefit/risk decision. Costs may include membership fees and travel expenses but will always include the time staff spends working on the standard that is taken away from other work. Even when the standard is being developed without face-to-face meetings in remote places of the world, staff will need to develop a plan for participations (goals, strategies, etc.), to take time to read, understand, and comment on specification drafts and to participate in phone meetings to discuss those drafts.

If one of your goals is to get one or more of your patent claims in the standard, additional time will be required to propose a technical feature for the specification. Additional time may also be needed to develop that feature. SDOs often have a procedure for making technical contributions. Staff will need to know that procedure and make your contribution in accordance with it.

If your organization cares about the content of the specification, you will need other SDO members to support your position. This may be more or less difficult depending upon the positions and power of other members. Your SDO representative will want to seek such support, unless it is happening on its own, but your organization's management may also want to reach out to the management at other member organizations.

Participating may require additional expenditures. The more formal the SDO the more likely that membership requires the payment of a fee. Larger, more formal SDOs frequently have multiple levels of membership with differing rights and fees. For example, the Institute of Electrical and Electronics Engineers (IEEE) has two levels of corporate standards membership, Basic and Advanced. Basic Members can attend and observe at standards working groups, but only Advanced Members can participate and vote.

Working groups often meet face-to-face to discuss drafts and ideas as well as having phone and email exchanges. This results in travel expenses and additional time away from the office.

Another key concern should be the impact of participation on your patent position. As noted earlier, participation in developing the standard generally requires you to license your patent claims in the mandatory or alternative portions of the standard. You may not want to license these claims to a competitor, especially if they cover a key differentiating feature of your product. Or you may be willing to license the claim for a fee, but the organization requires you to license to implementers without charging a fee.

This may not be a problem when you initially join an SDO or work group, but may become a problem later. The initial scope of work for the group may be such that you do not have an essential patent. In some SDOs the scope of the standard may change during the development of the specification. You will want to be alert for such potential changes. In addition, a patent may or may not be essential depending on how the specification is actually written.

It is possible that as the SDO's work goes on you change your mind about your willingness to license your essential claims and want to consider leaving the organization to avoid this obligation. The intellectual property rights (IPR) policies of most SDOs contain provisions covering the withdrawal of members from a working group or the organization. The licensing consequences of withdrawing vary from group to group and should be understood and evaluated when evaluating the risks of participating in a new standards development activity prior to joining.

STANDARDS DEVELOPMENT ORGANIZATIONS

A wide variety of groups write standards. These groups may be referred to as standards development organizations (SDOs), standard setting organizations (SSOs), standard setting bodies (SSBs), or by other terms. Some are formed to develop a single specification and then disappear; others continue to be relevant for many years. The ITU was founded as the *International Telegraph Union* in 1865 and continues to function as one of the world's most important standards development organizations.

Each group must make choices on a many organizational points such as what it will and will not work on, membership, organizational structure, membership fees, decision-making process, patents, and copyrights. For example, members in SDOs may be individual, companies, countries, , or some combination of these.

These choices are usually reflected in the SDO's written documents (articles of incorporation, if incorporated, by-laws, membership agreement, etc.). In some organizations, however, the actual practice diverges from the written rules. For example, there may be a written rule stating that specifications are adopted through majority voting, but the organization does not act unless unanimity is reached. Other SDOs take the position on key choices that "we will decide when we reach that point." Before joining it is best to check whether the organization's behavior conforms to its written rules.

HOW SDOS DEVELOP STANDARDS[1]

Standards development begins when an individual, company, or organization thinks that a specification or set of specifications is worth standardizing. It may have an idea for the standard or may have already done a considerable amount of work on the specification. At that point it will usually talk with other parties to see if there is interest in developing the standard.

If there is interest in joint development, the parties will discuss the best way to do this work. They may decide to work together under a new or existing agreement, form a joint venture, or work at a standards organization. If the work is to be done at an SDO, they must decide whether this work can be successfully done at an existing organization. SDOs vary widely in many important characteristics such as the speed with which they act, who can participate, IPR policies, and the credibility of a standard developed by the organization. Standards in some technical areas have no credibility unless they are developed by a specific SDO. If there is no SDO meeting their requirements, they must decide whether to form a new SDO.

In more formal SDOs, the sponsor may need to bring a formal proposal to begin a new working group and write a new standard to a management board for a vote or to the organization's director; however in less formal SDOs this process can be much more infor-

1. Much, but not all, standardization occurs at the behest of companies. For the sake of simplicity, this paper will proceed as if that is true in the matter being discussed.

mal. Some international SDOs require that before it will consider a technology for standardization, it must have first been standardized at a lower-level organization.

The working group may begin by defining parameters or requirements for the standard or that may have been done as part of the approval process for forming the group. Once that is done the group will begin by discussing and producing a draft specification. Such submissions are usually called contributions. SDOs vary widely in what constitutes a contribution. In some SDOs submissions may only be made in writing on the SDO's form; in others, verbal statements made in a workgroup or email comments may be binding as a submission. Participants need to understand the SDO's definition of a submission before beginning to participate because making a submission may bind the participants and their employers or sponsors to additional legal obligations. SDO rules usually provide that contributors grant the SDO a copyright license in their contributions if the contributor owns the copyright. If any other party owns the copyright, the SDO may require that the contributor to provide a license from the copyright owner. A contributor may also lose its right to deny a license to its essential claims upon leaving an SDO if those claims were in its contribution.

A draft may also include by reference one or more technologies already standardized and add to it.

Some groups that call themselves SDO develop profiles. Wikipedia notes that "A profile consists of an agreed-upon subset and interpretation of a specification. Many complex technical specifications have many optional features, such that two conforming implementations may not inter-operate due to choosing different sets of optional features to support. Even when no formal optional features exist within a standard, vendors will often fail to implement (or fail to implement correctly) functionality from the standard which they view as unimportant… Also, some writers of standards sometimes produce vague or ambiguous specifications, often unintentionally, but sometimes by intention. The use of profiles can enforce one possible interpretation."

The draft will be distributed within the working group. There will be an opportunity for members to discuss and comment on the draft, and to recommend changes either in concept or in new language for the next draft specification. Again, some or all of these actions will be considered as submissions and come under the SDO's obligations applicable to submissions. Based on the group's decision-making procedures, a new draft will be produced and the review process will be repeated.

The rules of most SDOs provide that discussions and comments within the standards development process are confidential. Practically, however, the larger the working group and SDO the less likely conversations and suggestions are to stay secret. If it is important that your participants keep information, such as products in development, confidential, it is best to remind them of this periodically. Contributions are rarely subject to confidentiality requirements.

At some point the group usually concludes that the draft is near completion. The draft may be circulated to a broader group before a final vote is taken. For example, the World Wide Web Consortium (W3C) routinely distributes Working Drafts for "review by the community, including W3C Members, the public, and other technical organizations." Unless the SDO's rules so provide, there is no requirement that a draft specification be distributed for comment outside the working group developing it, however. Commenters who are not members of the working group may be required to sign a feedback license or similar document as a condition of submitting a comment. This is intended to ensure that the organization and implementers will be protected from a law suit if a comment made by a non-member is incorporated into the standard.

Eventually the hope is that the working group agrees that the draft specification is complete and in final form. Depending on the SDO, the working group may have the final vote on adoption of the draft or it may also need the approval of the SDO's membership or board. Each SDO establishes how many votes a draft needs to be adopted. Some SDOs require only a majority vote; others require a consensus, which is often interpreted as unanimity or no strong dissent. Some SDOs have categories of members and set requirements for each category.

Some SDOs or working groups develop one or a set of standards and then cease. Some SDOs and working groups have ongoing activity. The body may find that the standard contain errors and correct them, or it may decide to develop an improved version of the standard such as USB 2.0 or HTML 4.

SDO MEMBERS' OBLIGATIONS REGARDING THEIR PATENTS

Much of the standards process is shaped by two linked sets of interests: the desires of those writing the specification to create a standard meeting their needs, and the desires of those who want to see the standard implemented or to implement it themselves. There is an inherent tension between patents and standards. "[T]he objective of SSOs, which often consist of companies interested in the development of the technology in question, is to establish standardized technology that can be used as widely as possible." Most standards include patented technologies, however. Patents grant their owners the exclusive right to practice a technology and the right to file legal actions against others who practice the technology without the patent holder's permission. Thus, if the patented technologies in a standard cannot be licensed on terms acceptable to implementers, the standard will not be used in products and the standard has little utility.

To address the patent issue, most SDOs include provisions in their organizational documents to ensure that their members' essential patented technologies in the standards are available for license. These rules are usually called the SDO's Patent Policy or IPR (intellectual property rights) Policy.

The IEEE definition of an essential patent claim is typical. It defines such a claim as "any Patent Claim the use of which was necessary to create a compliant implementation of either mandatory or optional portions of the normative clauses of the [Proposed] IEEE Standard when, at the time of the [Proposed] IEEE Standard's approval, there was no commercially and technically feasible non-infringing alternative. An Essential Patent Claim does not include any Patent Claim that was essential only for Enabling Technology or any claim other than that set forth above even if contained in the same patent as the Essential Patent Claim. This obligation runs to members of the SDO but also non-members including partners and competitors [that implement the standard] alike."

SDOs also foster the availability of patented technologies for implementers by including in their IPR Policies provisions requiring working group members to disclose their essential patent claims in specification drafts and in their contributions to the specification drafts, and to agree to license their essential claims in the adopted standard.

Most SDOs require working group members to disclose whether they or their employer have essential claims, if the member knows. This duty to disclose usually runs only to participants' and their employer's patents, but the requirement can also be broader: it can extend even to participants' knowledge of the patents of others. This duty usually applies to patent applications as well as issued patents. Courts have held that the failure to disclose, when obligated to do so, acts as an implied license to those wanting to implement the standard. Therefore, it is important that participants understand their obligation and that of their employer, if any.

This knowledge potentially enables an SDO to determine whether the essential claims will be available for licensing and to modify the specification draft if there is uncertainty. For example, APCO 25 noted that it "only selected technologies which were covered by IPR when in their opinion...the owners of the IPR offered them in accordance with the Project 25 Memorandum of Understanding at fair and reasonable terms and conditions."

Almost all SDOs require working group members to license their essential patent claims in the final standard. SDO documents usually indicate the broad outline of licensing parameters. The most common parameters will be described in more detail below. The duty to license is limited to compliant implementations of the standard.

Some SDOs permit members to include their licensing terms for their essential claims if the final standard includes those claims.

There are a variety of ways for SDOs to address this issue in addition to requiring members to grant licenses. SDOs membership documents may contain patent license grants or otherwise make patents unenforceable, at least in certain contexts.

PATENT LICENSING REQUIREMENTS

Those SDOs that do not seek to prohibit the inclusions of patented technologies subject to patent licensing in their standards usually limit patent licenses to one of three types:

- Reasonable and Nondiscriminatory (RAND) or Fair, Reasonable, and Nondiscriminatory (FRAND)
- Reasonable and Nondiscriminatory with zero payment (RANDZ or RAND Zero)
- Royalty Free (RF)

RAND licensing is probably the most common licensing regime. GSM, MPEG-3, Blu-ray are examples of RAND standards. RAND licensing envisions a licensing fee and other terms as well.

There is very little case law on the meaning of the words "reasonable" or "nondiscriminatory" in the context of standards. Thus, what constitutes reasonable and nondiscriminatory terms is open to a wide range of interpretations. Fees may be per unit or there may be an annual fee. There may be fees on part of the ecosystem but not others. Fees may constitute a large part of a product's selling price or none of it. Today, it is not unusual for patent royalties to account for 30 percent of a smartphone's cost because of the intellectual property associated with camera and the video camera technologies along with other features.

A situation may arise that a fee that is seen as reasonable when a product is new and the per unit cost is high, but is seen by some as unreasonable later when the product becomes common place and the selling price of the product falls substantially. For example, when the cost of the key patents in DVD players (about $20 per unit) remained the same as the selling price of the player fell from $100+ to $35 many manufacturers felt squeezed.

While much of the discussion of licensing terms centers on license fees, there are likely to also be other terms of importance. Some licenses run a hundred pages or more. Other license terms may define: what can and cannot be done with the technology; distribution restrictions; potential liability; suspension of the license if you sue the grantor for any related patent claims; conditions that must be included in any agreement you sign transferring control of your essential claims; limits on sublicensing; use of other technologies; testing of final products; notice provisions; arbitration and other dispute resolution steps; and termination rules. Courts have recognized this. For example, the U.S. District Court for Northern California found that violation of the copyright in code distributed free of cost under an open source license may be eligible for damages under copyright law.

A RAND license policy may slow or even halt adoption of the standard or lead to market fragmentation if implementers are unwilling to pay a fee, especially if the fee is expensive relative to the product's sale price. However, at times important patent owners will only agree to participate in development of the standard and the licensing of their patented technologies if they are able to charge a fee.

RANDZ licensing is also a very common licensing regime. Conceptually, RANDZ licensing is the same as RAND licensing except with a prohibition against charging a fee for licensing essential patent claims. Some observers argue that documents containing RANDZ standards should also be available free of charge, i.e. there should be no copyright fee for accessing the actual specification document.

If charging a fee for the use of patented technologies discourages implementation, eliminating the fee must encourage standards participation, adoption and market consistency. Whether this is a good thing depends on the business and patent position of your company.

There is no generalized agreement on what this means. Some SDOs probably equate the term with a RANDZ policy. Others may mean that a patent holder must agree to not enforce its patent claims against implementers. Because of this lack of clarity, extra caution should be used when considering participation in any SDO with such an IPR policy.

The patent policies of most SDOs include provisions defining participants' ongoing licensing obligations if they leave the working group or membership in the organization. These provisions vary widely and it is common for participants to have continuing licensing obligations even after they have terminated their memberships.

OTHER SDO APPROACHES TO PATENTED TECHNOLOGIES

There are several other approaches to handling patented technologies in addition to required licensing. A few organizations, such as the Open Web Foundation (OWF), have sought to simplify their approach to patent and copyright protection. The OWF Foundation Agreement requires signers to provide patent non-asserts. A non-assert is a promise to not seek enforcement of rights under one's patents against a party using the essential claims in that patent to implement the covered specification subject to stated conditions. This agreement also obligates signers to provide RANDZ licenses if requested.

A few SDOs attempt to create standards that do not include patented technologies. Meeting this goal can be difficult. As noted already, many of the technology members want to include in a standard to make it work well may already be the subject of patent applications or issued patents. In addition, patent law discourages patent searches because a search may serve as evidence that an entity engaged in a knowing, willful, infringement, rather than inadvertent infringement, resulting in greater damages.

Finally, some organizations allow patents but only if the contributor agrees to allow implementers to use his or her essential patent claims without the implementer getting a license and if the implementer makes the same agreement for its patents.

REQUIRED PATENT TREATMENT OF ESSENTIAL CLAIMS AND YOUR COMPANY

It is crucial to understand an SDO's handling of essential claims, your patent holding, and your business plan before joining any SDO. If you do not have a patent or patent application within the scope of the SDO's work, this is a non-issue. You must give the situation more thought if one of your patents is likely to include an essential claim in a specification. Does the patent cover a key part of your current business or the intended future direction of your business, or does it cover a technology you developed or acquired but will never use? If the patent includes a key part of your current or future business, will licensing it to others be a great way to collect revenue or will it allow competitors to implement at no cost a feature that currently enables you to differentiate your product? If your company is of a substantial size, you may need to check with other business units to determine whether a patent is important in their business plans.

Part of this analysis should include an understanding of the continuing legal obligations you will have if you leave the working group or SDO. Some SDOs allow members to leave their organization and totally avoid the member's duty to license their essential claims, but in most SDOs there is some degree of continuing obligation. In some SDOs if you terminate your membership, your licensing obligation runs not just to specifications adopted during your membership but continues in subsequent versions of the specification developed after you have gone. You need to understand these rules for the particular SDO in advance of joining.

This judgment regarding patents must be made on a case by case basis with input from both business and legal staff. In no circumstance should this analysis be skipped.

MANY SDOs DO MORE THAN JUST DEVELOP STANDARDS

SDOs develop standards. Some distribute copies of their standards for free; others charge. Some SDOs do considerably more. Here are a few examples:

- Develop tests to determine compliance with a standard
- Assist with implementing the standard
- Market the standard
- Test interoperability

DEVELOP TESTS TO DETERMINE COMPLIANCE WITH A STANDARD

It is common for groups developing a specification to also develop one or more tests to determine whether an implementation meets the requirements of the specification. These tests often take the form of a series of inputs with corresponding outputs. After a company finishes its implementation, it can use these tests to evaluate its implementation. If the implementation passes the tests, the company may claim its product is compliant with the standard.

Importantly, some standards are such that two products may both pass the compliance test and not work with each other. For this reason, some industries use plug fests or third-party testing to determine interoperability.

The Wi-Fi Testing and Certification programming is an example of a third-party interoperability testing. According to the program's sponsor, the Wi-Fi Alliance:

> Wi-Fi CERTIFIED is a program for testing products to the 802.11 industry standards for interoperability, security, easy installation, and reliability. The Wi-Fi CERTIFIED logo is an assurance that the Wi-Fi Alliance has tested a product in numerous configurations and with a diverse sampling of other devices to ensure compatibility with other Wi-Fi CERTIFIED equipment that operates in the same frequency band. The Wi-Fi Alliance network of independent test labs conducts interoperability testing programs to ensure that wireless devices work together and support secure connections.

Because of the potentially adverse effects from failing such tests, testing agencies only release the names of products that pass. Some also assist implementers in modifying their products so they are able to pass.

ASSISTANCE WITH IMPLEMENTING THE STANDARD

Some SDOs provide assistance for potential implementers. One way they do this is by requiring a reference implementation as a condition of starting a new standards development activity or soliciting a reference implementation from the members. A reference implementation is an example of an implementation showing that an implementation is possible. For competitive reasons, a reference implementation is not intended to be the "best" implementation nor is it intended to contain additional features or functions; these are reserved to differentiate products.

MARKET THE STANDARD

Some SDOs take affirmative steps to market their standards. This marketing can take many forms including press releases, articles, participation in trade shows, demonstrations, press tours, analyst briefings, or a logo program discussed above.

One example is the OASIS OpenDocument Format (ODF) Adoption TC. The TC describes itself as follows:

> The purpose…is to create awareness and demand for a new class of applications and solutions designed specifically to support and leverage OpenDocument XML. The Adoption Committee will dedicate its energy and resources to create wide scale understanding of the benefits of OpenDocument format support within organizations and governmental bodies through education and promotion. The Adoption Committee will align and support the activities of the OASIS OpenDocument Technical Committee by providing market-based requirements. These requirements will help guide future development of the OpenDocument specification by the OASIS OpenDocument Technical Committee.
>
> By advancing the adoption of OpenDocument within OASIS, the Committee members:
>
> - collaborate within the security of the OASIS open process;
> - maximize message credibility by working on behalf of a vendor-neutral standards consortium;
> - build on existing market awareness of OASIS as the home of OpenDocument;
> - leverage the combined resources of all participants;
> - coordinate closely with the existing OASIS OpenDocument TC development team;

- reinforce OpenDocument as the result of an open, transparent process guided by multiple vendors, communities and individuals; and
- openly share content, and coordinate program execution with other entities that share a common or similar statement of purpose on a worldwide scope.

Working together, members of the OASIS OpenDocument Format Adoption Committee will provide expertise and resources to educate the marketplace on the value of the OpenDocument format, increasing the demand for ODF-conforming products, and expanding the OpenDocument community of users, suppliers, and developers.

With the support of OASIS staff and in accordance with the OASIS Media Relations Policy, the Committee will assist in coordinating press and analyst relations such as briefings, releases, and announcements. The ODF Adoption TC will coordinate communications to the OpenDocument community through the OASIS and XML.org websites, collateral, newsletters, webinars, conferences, and other outreach venues.

In cooperation with the OASIS OpenDocument TC, the OASIS OpenDocument Format Adoption Committee may create white papers, FAQs, overview documents, case studies, user scenarios, and other educational materials. The Committee also may establish liaisons in accordance with the OASIS Liaison Policy with other organizations, government agencies, and academic institutes concerned with office applications to broaden awareness of OpenDocument. The Committee may also assist in coordinating the promotional efforts of members relating to OpenDocument, leveraging those activities wherever appropriate. The Committee may work on proof-of-concept demonstrations to showcase how the

> OpenDocument Format can be integrated with other standards, for example using ODF, XFORMS, UBL and various Web service standards built using specific industry vocabularies, schemas and/or transaction sets.
>
> This Committee consists of volunteers from member organizations and will conduct teleconferences and occasional face-to-face meetings.

Some SDOs also adopt trademarks. If you will want to use such a trademark you will need to understand who owns it and the requirements for and restrictions on its use so you will be able to determine whether these conditions align with your plans.

ANTITRUST LAW AND STANDARDS DEVELOPMENT

Because most standards development activities include concerted action among market competitors, participants must take care so they do not violate antitrust laws. In addition, some SDOs have antitrust policies. The INCITS Antitrust Guidelines are typical. Among its provisions is a list of "Sensitive Topics" that "[w]ith rare exceptions that should be made only upon the advice of INCITS counsel, there should never be discussion of the following topics at any INCITS or an INCITS subgroup meeting:

- Any company's prices or pricing policies;
- Specific R&D, sales and marketing plans;
- Any company's confidential product, product development or production strategies;
- Whether certain suppliers or customers will be served;
- Prices paid to input sources; or
- Complaints about individual firms or other actions that might tend to hinder a competitor in any market.

Participants should discuss this topic with legal counsel before beginning their participation.

IMPLEMENTATION OF STANDARDS

As noted above, a standard that is not implemented is, for most participants, a significant waste of resources. To help encourage potential implementers, the standards world has developed several actions including disclosure requirements, mandatory licensing provisions, and reference implementations discussed above as well as patent pools and open source concepts.

LICENSING

LICENSING IN GENERAL

In requiring disclosures and listing the patents disclosed, SDOs give potential implementer a list of patents to review and patent holders to contact. Most patent standards licensing is a one-to-one transaction: a potential implementer approaches a patent holder to seek a license or, less commonly, vice versa. There are negotiations and an agreement may be reached.

Occasionally, the parties may already have a cross-license in place that covers one or more of the essential claims.

There is no guarantee that the disclosure list covers all of the essential claims in the standard. It may be that the holders of some of the essential patent claims are not members of the SDO or the working group. Also, a participant may be unaware of key patents held by his or her company. Thus, it is possible for an essential patent claim holder to surface after the standard is adopted or implemented.

PATENT POOLS

A patent pool is a legal arrangement in which two or more holders of essential patent claims agree to jointly license these claims under a single license. The formation of a patent pool usually begins with one or more licensing groups putting out a solicitation asking holders of potentially essential patents to submit these patents to the organization for evaluation if the holder is interested in potentially participating in a pool. Patent holders need not have been a member of the SDO or working group that wrote the standard to submit a candidate patent. Nor is there a requirement that everyone who participated in the SDO and has an essential patent claim must submit its claim or participate in a pool.

The group hires an independent expert "to determine which patents are essential..., thereby defining a group of essential patent holders." These patent holders then attempt to agree on a license. "A patent pool administrator appointed by the essential patent holders to handle administrative tasks such as signing up licensees, collecting royalties from the licensees, and distributing the royalties to the essential patent holders...The essential patent holders retain...the right to license the patents outside of the patent pool."

A well run patent pool can assist both implementers and licensors. A patent pool can lessen implementers' licensing transaction costs. By enabling licensees to sign one license and pay one fee covering multiple patents, it lowers the number of negotiations that must be conducted and the number of fees that must be paid. Note that here, too, there is no guarantee that a pool covers all of the essential claim holders in the standard.

Patent pools can also benefit patent holders for much the same reason. Once the pool license is agreed to, the pool administrator deals with potential licensees and with fee collection. Even so, some companies are reluctant to join patent pools.

OPEN SOURCE SOFTWARE

Some commentators feel that charging any fee is unfair because they believe "Patent royalties tend to create discrimination against small-to-medium-sized businesses developing any form of software, and especially against Open Source developers."

"Open source software is built upon the principle that end users should be given source code and should be free to use, share, modify and enhance software products, with the goal of widespread interoperability and permissive incorporation into new technology." "Open-source software (OSS) is computer software that is available in source code form for which the source code and certain other rights normally reserved for copyright holders are provided under a software license that permits users to study, change, and improve the software."

Open source licensing focuses on copyrights and software implementations rather than patents. "Authors initially derive a right to grant a license to their work based on the legal theory that upon creation of a work the author owns the copyright in that work. What the author/licensor is granting when they grant a license to copy, modify and redistribute their work is the right to use the author's copyrights. The author still retains ownership of those copyrights; the licensee simply is allowed to use those rights, as granted in the license, so long as they maintain the obligations of the license. The author does have the option to sell/assign, versus license, their exclusive right to the copyrights to their work; where upon the new owner/assignee controls the copyrights. The ownership of the copyright (the "rights") is separate and distinct from the ownership of the work (the "thing")—a person can own a copy of a piece of code (or a copy of a book) without the rights to copy, modify or redistribute copies of it."

"When an author contributes code to an Open Source project (e.g., Apache.org) they do so under an explicit license (e.g., the Apache Contributor License Agreement) or an implicit license (e.g., the Open Source license under which the project is already licensing code). Some Open Source projects do not take contributed code under a license, but actually require (joint) assignment of the author's copyright in order to accept code contributions into the project (e.g., OpenOffice.org and its Joint Copyright Assignment agreement).

Placing code (or content) in the public domain is a way of waiving an author's (or owner's) copyrights in that work. No license is granted, and none is needed, to copy, modify or redistribute a work in the public domain.

One author has estimated that "there are close to 2000 different open source software licenses (either self proclaimed or certified) in circulation. [There is no reason to] expect each of them to carry exactly the same terms and conditions." Examples of free software license / open source licenses include Apache License, BSD license, GNU General Public License, GNU Lesser General Public License, MIT License, Eclipse Public License and Mozilla Public License." He recommends "that each OSS component has to be looked at and analyzed individually before one can reasonably determine the implication of its use in a given product."

Many open source advocates believe that implementers should not be required to get patent licenses as a condition of implementing a standard. The Open Source Initiative states, for example: "The purpose of an open standard is to increase the market for a technology by enabling potential consumers or suppliers of that technology to invest in it without having to either pay monopoly rent or fear litigation on trade secret, copyright, patent, or trademark causes of action. No standard can properly be described as 'open' except to the extent it achieves these goals."

Bruce Perens is a leading Open Source advocate. According to Wikipedia, "Under Perens' definition, open source describes a broad general type of software license that makes source code available to the general public with relaxed or non-existent copyright restrictions. The principles, as stated, say absolutely nothing about trademark or patent use…"

A key difference between many commercial products and open source products is that most commercial products include warranties that protect buyers in the event of a patent infringement claim. Each business must decide for itself the importance of such a provision.

CONCLUSION

Standards have become increasingly important as businesses and government work to make their resources stretch farther because standards have the potential to increase productivity or allow products to be reused rather than replaced. This has been reflected in two types of purchasing preferences: the requirement that the desired product comply with one or more specific standards; or the requirement that the product be based on an open standard rather than a proprietary technology. In this business environment the question of whether to participate in standards development is a natural one.

Standards participation can be a high risk high reward endeavor. In addition to the resources required for participation, patent holders may become obligated to license the technology that forms the foundation of their business to competitors for free. Conversely, standards participation may result in international demand for your technology that would have been difficult to generate any other way.

It is essential to determine how the SDO works before you decide whether to join it. If the SDO will be formed, it is essential to know how it will work. Determining how an SDO works is first a matter for legal counsel. If the SDO is already formed, counsel should analyze its legal documents. This analysis should be confirmed by someone with experience in the organization. If you do not have experience with the organization in-house, so you can understand its customs and unwritten rules, you may want to reach out to a friend at another business or to the SDO itself. Once that is done, counsel and business staff should meet to determine your goals for participation, a plan for how to reach you participation goals, the likely impact of participation on your business, how that fits with your business plans, and whether participation or non-participation will have any publicity, partnership, regulatory or other impacts.

If a new SDO is being formed, counsel and business staff should meet to determine the same matters discussed above but also need to ask whether you need the SDO to have a certain structure to meet your goals. If the answer to this question is yes, you must then decide if you are going to try and influence that structure or your own or with partners.

CHAPTER 8

A Patent Judge's Perspective

Luann Battersby

Luann Battersby is from the diminishing club that hails from an agricultural/farming background. She has spent nearly all of her three-decade career as a scientist, first as a food scientist (B.S. Penn State) working in the food industry (Kelloggs, ITT Continetal Baking, others). Later as a molecular biologist (M.S. Hood College, M.H.S. Johns Hopkins School of Hygiene and Public Health). Ms. Battersby worked for USAMRIID in infectious disease biocontainment research, wearing a moonsuit regularly. She progressed to biopharmaceutical research and clinical trial design and laboratory testing at MedImmune and Human Genome Sciences, among others. Heading to law school later in life (Georgetown J.D.), Ms. Battersby combines scientific understanding with law bringing valuable insights to some of the most compelling bio-technological and software patent litigations of our day. Ms. Battersby is a licensed attorney before the USPTO, the State of Maryland, the District of Columbia, and the Federal Circuit Court of Appeals. Ms. Battersby can be contacted at Luann_Battersby@comcast.net.

If a patent examiner maintains a final rejection, one option is to take the case up on appeal. Luann Battersby, a patent attorney in Pennsylvania, interviewed Neal E. Abrams, a retired Patent Office appeals judge to get his perspective and advice for applicants. She has included some of Mr. Abrams' personal background to provide a better sense of who he is, and thus what type of person may be looking at your appeal.

MR. ABRAMS' CAREER

Never imagining that law might be his eventual career, Neal E. Abrams found the job of an appeals judge for the USPTO to be the best of his life, and one in which he made a difference. Retired Judge

Abrams describes his ascent to that position as being quite a typical pathway, at least at that time. He participated in 5000 cases, writing 1800 opinions, or over 100 opinions per year. None of his opinions were ever reversed. His is a fascinating story which was far from predictable at any point. Mr. Abrams offers his insider advice to those who wish to be successful before the Patent Office Board of Appeals.

Reviewing the career of Neal E. Abrams, who maintained his judgeship with the Patent Office Board of Appeals for 17 years, it becomes obvious that technical excellence and hard work were the cornerstones of his successes. Mr. Abrams did not begin with such a lofty goal in mind. Following the University of South Carolina and ROTC and a degree in mechanical engineering, he joined the Navy and was assigned to the Guided Missile School. Upon leaving the Navy, Abrams won a job with the U.S. Patent and Trademark Office (USPTO) making the princely sum of $2800 per year as a patent examiner in an area that would not only dovetail with his love of cars, but which would also be an underpinning to the rest of his professional life: he examined in the art area of Exhaust Systems for Internal Combustion Engines. Finding himself amidst engineers who shared his love of all things automotive, he drives a Corvette racecar today. Mr. Abrams followed the advice of a York, Pennsylvania, patent attorney and headed to American University for four years of night law school.

Within his first two years of night law school and USPTO employment, Abrams recognized that the USPTO employed two types of people: the lifers, and those who were destined to find their way out into industry or other branches of government. He joined the departing second group as a patent advisor to the Navy Bureau of Weapons, moving about every two years to jobs of increasing complexity and opportunity, from Atlantic Research Corporation in Fairfax, Virginia, to Budd, in Philadelphia, Pennsylvania, the manufacturers of the high speed Metroliner trains, until he finally came to rest for some years at a Connecticut and K Street law firm in Washington DC, where he shortly gained a partnership position. During this time Abrams prosecuted his most important patent application, one we all will be very familiar with: the plastic grocery cart. In 1977 Abrams moved to the Department of Energy and used his skills in the area of alternative energy sources including windmill farms, wave motors and methane from biomasses such as seaweed and even DC waste. He then served with the Nuclear Regulatory Commission from 1980-1987 as Chief Patent Counsel, even acting as a labor lawyer before returning to the USPTO and becoming an Administrative Patent Judge on the Patent Office Board of Appeals in the Mechanical Arts area where he would remain for the next 17 years. The climb through technologies and industries with a stint on K Street by a man with a chemistry or an engineering background was typical for judges of those days, says Abrams. Judges are grouped into chemical, electrical or mechanical disciplines.

Abrams found highly motivated and bright people throughout the process. He enjoyed being in the position of judge for the Board of Appeals for many reasons. In many governmental agencies, the Administrative Law Judges (ALJ) render opinions that are advisory. That is not the case with the USPTO Board of Appeals judges; those decisions are real, says Abrams, "This is a no-B.S. job. The final decision of the Board of Appeals represents the USPTO's opinion on patentability."

TIPS ON THE PROCESS

Some background on how a patent application gets to the USPTO Board of Appeals is in order. After a patent examiner's final rejection a patent applicant may request to make an amendment after final confirmance. If the amendment does not result in patent issuance, there is a route of appeal for the patent applicant/appellant by which the appellant files a notice of appeal and brief. Although this route is within the USPTO, the USPTO Board of Appeals operates quite independently of the USPTO's patent examiners, deciding the cases as the judges see fit. During Abrams's tenure, about 50-65% of the patent office's examiner decisions in the Mechanical Arts area were reversed by the Administrative Law Judges of the USPTO Board of Appeals. These USPTO Board of Appeals *inter partes* proceedings are typically decided on the record with only about 5% of cases heard live, a choice that is made by the patent applicant. Live appeals are oral arguments limited to 30 minutes. There is one chance to argue and the attorney cannot argue outside the brief. The patent examiner can appear if they wish to, but that is the exception. Whether on the record or oral, three judges review each appeal of an examiner's opinion, one judge is assigned to write the opinion, this pivotal judge confers with the second judge, and a third judge is available if there is spirited disagreement between the writing judge and the conferring judges. The judge sitting in the center is the senior judge who will write the opinion. While the judges are not robed in the USPTO Board of Appeals, the atmosphere is formal and the address is the traditional 'Your Honor.'

If the outcome of the case is a loss before the USPTO Board of Appeals, an applicant can ask for reconsideration by the Board of Appeals. It may be granted but the outcome is unlikely to change. The decisions of the USPTO Board of Appeals are further appealable by two routes. The most common route is by appeal to the Court of Appeals for the Federal Circuit (CAFC), a court well-accustomed to patent issues, which, itself, can be appealed only to the U.S. Supreme Court at their discretion. An alternative appeals route is by appeal to the District of Columbia District Court, which can then be appealed to the Federal Circuit, before that court's decision can potentially be appealed to the U.S. Supreme Court.

Attorneys before the USPTO Board of Appeals tend to be Type A realists, notes Abrams. There are big issues involving technology and the law and large sums of money are typically involved if the appeal route is sought. Negotiate with the patent examiner claim-by-claim before heading for an appeal, advises Abrams. If someone wants to buy the invention for a fair price, with or without a patent, sell it! If you represent the patent applicant and there is disqualifying 'prior art not seen in the patent office' in discovery, settle! However, if negotiation with the examiner does not work, if the disqualifying prior art is open to argument or interpretation as to its putative disqualifying characteristics, and there is no good buyer without a patent, appeal!

Abrams advises some important unwritten rules for those who wish to be successful before the USPTO Board of Appeals:

- The basics—argue the facts and the law!

- Know the rules
- In writing a brief, there is always a temptation to show the client that one has done a lot of work on their behalf by writing exhaustively. Resist that temptation. This is not a time for wordiness. The judges are busy people with an enormous workload. Be right and get to the point.
- Never disparage the examiner; do not attack the USPTO, attack the references, either their content or their quantity—four is at least one too many.
- Take the examiner's comments apart, word-by-word and line-by-line looking for any flaw in the reasoning or weakness in the scientific underpinning. Find the deficiencies in the case and put them in the brief.
- Accept the correct points made by the examiner, gaining credibility for your own arguments.
- Focus on the claims; not what you think the claims are, not what you think the claims should be, and not what is in the specification. Your argument should be directed to your client's claims.
- Know the law—103 rejections are an area in flux. For example, has the KSR case regarding known teachings removed Teaching-Suggestion-Motivation as requirements for combining references or has it modified what Teaching-Suggestion-Motivation stands for slightly or instead changed the meaning of any of those terms? If this is a 103 rejection, what would or should a person of ordinary skill in the art have known at the time of the invention?
- There are really only two arguments: the prior art is missing a limitation, or there is no teaching and one of ordinary skill in the art would not have found the invention obvious.
- What about experts and expert opinions? Use them to bolster your brief.

- Treatment of less than all the claims by the examiner should make those unexamined claims allowable if nothing has been noted regarding them. File the appeal and argue all claims. If you do not argue all claims, it is presumed that you are lumping all the claims together. On appeal the judge can touch all the claims whether the examiner touched on them or not.
- If your client has opted for oral argument and you are not articulate and personable, find someone who is to do the oral argument and confine your efforts to knowing the law and the facts and writing the brief. Make sure the brief covers any elements that will be used in argument.
- In oral argument, point out the points you have made in your brief on the record. Use the visual technologies that are available to show the state of the art at the time of the invention and how your invention differs.

Even a rejection by an examiner or the Board of Appeals may be useful. "The examiner is doing the patent applicant a favor if they reject, if it's a valid rejection," declares Abrams noting that the money to enforce a bad patent will be wasted.

The 1990's and 2000's have seen many changes for Board of Appeals Administrative Patent Judges. There was a tremendous increase in patent applications—and hence in appeals—causing the number of judges to double from the 32 who Abrams worked with during much of his career. Patent attorneys began to be provided as law clerks, and the pay improved. Judges may now work primarily from home, a trend the USPTO has pioneered for its examiners as well. Most recently, webcam technology has allowed not only the writing of opinions from home but also the ability to hear 'live' appeals cases via the webcam linkup.

The Internet has changed the whole art of patent examination in that formerly patent claims were primarily rejected based upon other patents. The USPTO, through its international on-line subscriptions, now rejects claims based upon many different sources of knowledge and literature.

The very recent and long, long-awaited patent reform act, called the America Invents Act, has shifted the U.S. patent system at last from a litigation-spawning first-to-invent to a first-to-file system to harmonize with most other patent offices around the world. Judge Abrams dealt with the first-to-invent system which greatly complicated the system with such expensive legal battles as Interferences. While former Judge Abrams acknowledged that the U.S. patent system was broken in a few places and that the more streamlined first-to-file system will probably work better, he commented that leaving patent validity squabbles to be fought after a patent is shown to have value makes some sense. He notes that the U.S. inventor and U.S. small business is like no other in that their inspiration is often found in their garage workshop and that many, many inventions are still the product of one fertile mind. Although Judge Abrams had advocated for a change for 20 years by the time he retired in 2004, he wonders if changing the system to a first-to-file may discourage the efforts of the garage-one-man-show type of inventor.

Abrams suggests that more examiners are needed, that they need more time to make their determinations, and that the system needs to be simplified. Both examiners and judges seem to be inundated by the massive numbers of patent applications and their appeals. Patent applicants are reporting a three-year pendency and that the Board of Appeals is overwhelmed with a 22,000 case backlog even with keeping up the pace Abrams did, writing an opinion on two cases a week for nearly two decades.

Today former Judge Abrams is semi-retired to a small Pennsylvania town where he is well-known for his prodigious and lengthy volunteer civic service and steadfast personal support of the local EMT and fire department where he serves in a leadership EMS role and is also a member of the Board of Directors. He continues to privately consult on patent issues through an 'Of Counsel' position to a Pennsylvania law firm.

Section II

Valuation

CHAPTER 9

Patent Disputes

Eric M. Fudo

Mr. Fudo is a Director with BBP Partners, LLC. He consults extensively with attorneys and other clients on a variety of issues in litigation, arbitration, and other financial disputes. He performs financial analyses and guidance to parties involved in litigation and provides expert witness testimony regarding lost profits and economic damages issues. In addition, Mr. Fudo performs financial forensic investigations related to intellectual property licenses, alleged misappropriations and corporate fraud issues.

Mr. Fudo specializes in performing litigation and financial consulting engagements as well as forensic investigations and has been retained as an accounting and/or financial expert on many occasions. These projects generally include a detailed analysis and evaluation of accounting, financial and economic issues. He has experience in a wide range of industries including high-tech manufacturing, hospitality, financial services, industrial goods manufacturing, consumer goods manufacturing, electric and gas utilities, construction, entertainment, and transportation. He also has significant experience performing investigations in foreign countries, including China, Japan, Taiwan and several European countries. Mr. Fudo can be contacted at efudo@bbppartners.com.

To put "Patent Disputes" into context, let us elaborate a little bit. What we are talking about is the litigation, arbitration, mediation or negotiation between two or more parties concerning their respective rights, or lack of rights, associated with one or more of the parties' patents. My expertise however does not necessarily lie with the parties' rights, or whether there has been infringement. That is something for technical experts (engineers, scientists, programmers, etc.) and the attorneys to prove. My expertise as an accountant is in dealing with the economic consequences of one more of the parties' unauthorized use of patented technology(s). I am going to discuss how one would go about putting a price tag on the damages caused by an infringer.

In doing this I am going to attempt to keep things a bit simple. I am going to talk about disputes as though they are between two parties and as though they involve a single patent that represents a single technology. Real-world patent infringement disputes are seldom this simple. In fact, infringement allegations are often between parties that compete in the same industry, that sell similar products or services to similar customers through similar distribution channels in similar geographies. As such, disputes are often complicated by counter-allegations of patent infringement, client sensitivity issues, other trade secret complexities and the involvement of related parties.

Moving forward into the more technical portion of this chapter, it is important to keep several things in mind:

- First, I am an economic damages expert and not a patent owner or a patent infringer. As such, I'm not writing from a perspective of advocacy. I am writing as a person who is most often a neutral observer and evaluator of the available facts.
- Second, every matter and the associated fact patterns are unique, and strategies and theories employed in dealing with one matter are not necessarily relevant to another matter, however similar it may be. However, there are generally accepted means of dealing with the economics of patent infringement.
- Third, I may occasionally use the term "patented product," which is intended to be shorthand for "a product that incorporates the patented technology."
- And finally, much of the language below is addressed to the reader as though they are the patent owner, or the infringed. The concepts, however, apply to all parties involved in a patent dispute.

OVERVIEW

Regardless of the type of dispute you may be facing, and regardless of whether you are the plaintiff or the defendant, it is absolutely essential that you arm yourself with as much information pertinent to the economics of the dispute as you can. Some of this is going to be your own information, some of it will be your adversaries' information and some of it will be publically available information related to third party issues that deal with a comparable set of facts. Proving you have been harmed by some specified amount of money, or minimizing the amount of harm you have caused, is a complicated task and failing to consider all available information may have devastating consequences. Recent case law has made certain that successfully demonstrating the economics of your case will be wholly dependent upon the quality and reliability of the information you use, and the reasonableness of the assumptions you make to fill in the blanks. You will not be successful by merely asserting that there were damages.

With that said, *wanting* to gather pertinent information does not mean you will be able to do so. In litigation, arbitration and, to some extent, mediation, you will have the benefit of the discovery process. I will not get into the challenges of dealing with adverse parties through the discovery process, but you should know that it may not result in the data and information you need to prosecute your case. What do you do if you do not even have the benefit of court ordered discovery? Honestly, there is not much you can do short of making the best of what you have. The good news is that an adversary that does not produce information you deem relevant will not be able to rely on that information in court either—in general, check with your lawyer. An adversary that has knowledge of something you do not can still have an advantage. Either way, you will no doubt make use of lawyers and experts that are experienced in dealing with these issues.

CALCULATING PATENT INFRINGEMENT DAMAGES

There are a number of ways to value of your patents, but in a damages situation the involved parties will most likely be confronted with what is called a "but for" concept. That is, analyzing the difference between the position the company would have been in, but for the patent infringement, and their current position. To help illustrate this concept:

The income that should have been earned		$1,000,000
Less:	The income actually earned	- $250,000
Equals:	Damages	= $750,000

The before-and-after can take many forms. Looking at this from the perspective of the damaged party, you may be dealing with, among other things:

1. Actual profits (or losses) generated from patented products have been negatively impacted by an infringer;
2. Royalty rates from licensing the patented technology to third parties have been negatively impacted by an infringer;
3. The number of licensing agreements that have been put in place have been negatively impacted by an infringer; or,
4. The revenues and/or sales volumes your licensees are able to achieve, and thus the royalties you are paid, have been negatively impacted by an infringer.

Each of these can be caused by the unfair presence of the infringing competition and the associated pressures that that competition puts on the patent owner's and their licensees' ability to sell more product at more favorable pricing. Or, at least that's the claim. Proving it is another story.

LOST PROFITS

A claim for lost profits associated with product sales assumes, first and foremost, that the damaged party was in a position to make profits but for the infringement. So before we get into how to go about doing this, the infringed party must ask of itself four questions:

1. Was there demand for products containing the patented

technology during the period of infringement;

2. Was there an absence of acceptable, non-infringing substitute products that could have satisfied demand during the period of infringement;
3. Was the infringed party capable of meeting the demand of customers that purchased infringing products; and,
4. Can the profits that would have been earned be measured to a reasonable degree of certainty?

Being able to answer three of these in the affirmative is not good enough. In order to demonstrate that profits would have been earned but for the infringement, the courts will require that the infringed party prove each of these. That is not an easy thing to do for start-ups and companies that are not yet selling patented products. Actually, it is not even an easy thing to do for companies that are selling patented products. For example, imagine you are in the market with not just a competitor that infringes your patents, but also six other competitors that make non-infringing products that might be considered acceptable substitutes to some consumers. Of course, the patent owner will portray a narrow view of the market demonstrating that the patented features of their product render all other products inferior and poor candidates for substitution. The infringer will likely take a broader look at the market and identify many competitors. Absent the infringing party and their sales, is it reasonably probable that the patent owner would have made all of the infringer's sales? If not, is it reasonably probable that the patent owner would have made some of those sales? If yes, how many? Sometimes these questions are answerable, sometimes they are not.

Assuming the answer to each of the four questions above is yes, then what are the lost profits? To answer that, we need to go back to the previous discussion and determine what would have happened but for the infringement and create a hypothetical statement of earnings over the period of infringement. To do this, keep in mind that such an analysis need only be done to a reasonable degree of certainty; somewhere between unfounded assertions and absolute precision, but certainly closer to the 'absolute precision' end of the spectrum.

LOST REVENUE

So let us create a new statement of earnings, line by line, starting with revenue. This is relatively straightforward. We merely need to determine how many patented products would have been sold but for the infringement, and what the unit prices would have been. Right? Sort of.

To determine the product volume you will need to determine how many of the infringer's sales you would have recovered. And to do that, you need to know how many infringing products the infringer sold. Even absent other competitors, it is unreasonable to assume that you could have recovered 100% of those sales. There will be price elasticity issues (unless your pricing is lower), marketing differences, brand loyalties, product differentiators (maybe yours is not offered in pink), consumer exposure differences (the infringer's product could be convoyed from another of their very popular products), etc. You need to evaluate not just whether you *could* recover all their sales, but whether you *would* have recovered all their sales.

With that being said, there are instances in which you might expect to recover more than 100%. What if the infringer was first to market and had introduced a product of such low quality that its reputation for unreliability damaged the consumer's perception of the product category as a whole? Before you could get your product to market the irreversible harm had been done. In the "but for" situation, the harm may not have existed.

Another issue that needs to be evaluated is that the lost sales volumes are not limited to the past, and they are not necessarily limited to the patented product. Even assuming that as part of the court's judgment the infringer in enjoined from selling one more infringing product, that does not mean that the patent owner will be able to recapture those sales the day after the injunction. Unless you are dealing with some life supporting technology, there is most likely going to be a lag. Therefore, you will need to determine how long it will take for your actual sales catch up to your projected sales. It is a simplistic illustration, but take a look at Figure 1.

Assume your infringement damages calculation is being done in Year 5. If you assume that you will instantly jump to a sales level of 250 units per year, your damages calculation will only account

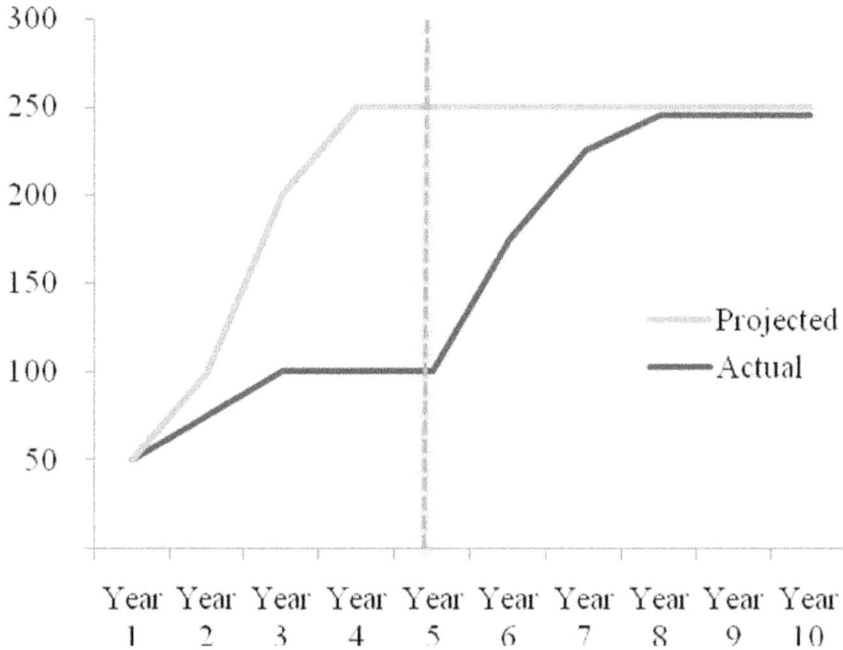

Figure 1: Hypothetical sales projections

for the lost sales represented by the area between the Actual and Projected lines to the left of the vertical line. However, if you assume that it will take some amount of time before you will be selling at a rate of 250 units per year (as we know actually happened because in our imaginary scenario we are looking at the actual results as of Year 10), you will also account for the areas between the Actual and Projected lines to the right of the vertical green line. You also need to recognize the possibility that you may never be able to achieve your original plans because of the actions of the infringer. In which case, damages may extend out many years, maybe even the entire life cycle of the patented product. In short, the period of damages is not limited to the past.

In addition to determining the lost profits associated with patented products, you need to also evaluate the possibility that other revenue may have been lost as a consequence of the infringement. Some of the more common forms of lost convoyed revenue include service contracts, replacement parts and product accessories.

The next part of the revenue equation is unit pricing, which was likely eroded as a result of the unfair introduction of the infringer to the marketplace. In the "but for" situation, however, that downward pressure on pricing would not have existed; at least not without other influences. So does that mean you can apply whatever pricing you want? After all, you would have had a monopoly. If you want to survive your opposition's and the court's scrutiny, pricing should be based on your historical pricing, ideally before any infringement, or possibly your infringer's pricing if you were not in the market first. But pricing is not necessarily a static number. If you were making a consumer electronic device, it would be unreasonable to assume that there would not be pricing declines over time. To the contrary, if your product cost relied heavily on an element known for upward cost volatility such as foreign currency translations, freight, or a meaningful amount of precious metal content, your pricing assumption might reflect this.

INCREMENTAL EXPENSES

The second part of the lost profits equation deals with the expenses that would have been incurred in generating the lost revenue. The key to this part of the analysis is determining the amount of incremental expense. It is not as simple as applying the net margin recognized during the period of damages to a new revenue number. On the other hand, incremental margins are likely to be something more complicated than the gross margin.

A typical approach in addressing the incremental cost structure is to analyze the damaged party's trial balance on a line-by-line basis. Each line would be evaluated in the context of the but for sales volume. The question asked of each expense category is whether the expenses are fixed, variable or some combination of the two. The characterization of expenses varies substantially from one industry to another, and even from one business to another within the same industry. The important concept to recognize is that you are trying to determine the *incremental* cost to add the additional units at issue. Is your CEO's salary going up? Maybe, if that salary is a function of profitability. Is your rent going up? Probably not, unless you need to add new facilities. Will cost of goods sold go up? Almost certainly. Will advertising expenses go up? Will utility consumption go up? Will freight go up? Will charitable donations go up? While the answers to many such questions might be highly subjective, having sufficient available data with which to perform a regression analysis could provide some necessary objectiveness.

Depending on the lost volume, even standard fixed costs may have to be revised. For example, if production output would have been 10 times more than the actual production experienced during the period of damages, but the manufacturing facility only had the capacity to produce half of that, you will have to take into account the cost of additional facilities. As an offset, excess volume may lead to lower per-unit raw material and direct labor costs.

Once you can define what the hypothetical income statement would have looked like over the period of damages, all that is left to do is compare that with what the actual income statement looks like. Keep in mind, of course, that each of these measures will likely incorporate a period that extends into the future.

But, what do you do if:

1. There is no demand for products containing the patented technology during the period of infringement;
2. There is an abundance of acceptable, non-infringing substitute products that could have satisfied demand during the period of infringement;
3. The infringed party was not capable of meeting the demand of customers that purchased infringing products; *or*,
4. Lost profits cannot be measured to a reasonable degree of certainty?

The courts have a long-standing answer. A damages award shall compensate the patent owner for the infringement, and in no event will that amount be less than a reasonable royalty. Translation: if you cannot prove that you have suffered lost profits damages, damages will equate to what a hypothetically negotiated license agreement between the patent owner and the infringer would have provided. And, your damages will not be less than what a reasonable royalty would have provided even if you can demonstrate lost profits.

REASONABLE ROYALTIES

What would result from such a hypothetical negotiation? Your first response is most likely, "I never would have negotiated a license with those jerks." And, that may be true. But in an effort to determine what a patent owner's damages are in the absence of lost profits, this has become a widely-accepted method of measuring those damages in the context of infringement. In order to make it work though, we will need to recognize the inherent differences from an actual negotiation by making a few assumptions:

1. Both parties were willing and able to negotiate a license;
2. Such a negotiation would have been at arm's length;
3. There was no threat of litigation;
4. Both parties recognize that the licensee's product infringes the patent;
5. Both parties recognize the nature of the benefits that the licensee derives from the license and the resulting sacrifice of the patent owner's profits, if any;
6. Such a negotiation would have taken place at the time of the first infringing sale; and,
7. Both parties would have shown their hands, so to speak, at the time of the negotiation.

Now that the framework for these hypothetical negotiations has been laid, we need to start negotiating. But, where do we start? At zero? It used to be that starting at 25% of net profits was the accepted norm. However, the courts recently ruled that starting hypothetical negotiations at 25% of net profits simply because it is "the rule" is arbitrary and fails to consider the facts of the particular situation. So, again, where do we start? Generally speaking, by establishing the value of the patent to the two parties using some combination of a cost approach, a market approach and/or an income approach as guidance.

The cost approach attempts to determine the value of intangible assets by calculating the cost to reconstruct or purchase non-infringing technology. This approach considers not only the cost of the "work around" or acquisition of the non-infringing technology, but also other risk factors like delayed market entry and inefficiencies associated with the alternative technologies. There are a few shortcomings with this approach, though. Primarily, the cost approach does not address either of the parties' expected benefits from the patent, nor does it address the probability of realizing those benefits.

The market approach looks to transactions involving technologies similar to those that are the subject of the hypothetical negotiation. First and foremost, you would look to what the patent owner and/or the infringer have done themselves. Have either entered into license agreements or transactions involving the same or similar technologies in the past? If so, what were the payment terms? In addition, you would look for publiclly available information related to what other unrelated parties may have negotiated. You would look for data associated with purchase transactions and data from other license agreements. As a word of caution, though, the courts are picky about such comparable information. To be used and relied upon as comparable, it must be comparable. You must consider, among other things, the relevance of the timing, geography, technology, company size, license terms and duration, the existence of cross-licensing, the relationship of the parties and whether terms were negotiated at arm's length.

The income approach focuses on estimating the present value of the profits derived by the licensee from the technology and splitting the profit between the licensee and the licensor. It is the amount of that split, of course, that is the central issue here. What this analysis does provide the user with is an understanding of the capacity to pay, or the requirement to be paid. For example, it is unlikely that licensor will license its technology for the chance of making $500,000 in royalties if it thinks it could make $1,000,000 by selling patented products on its own. Similarly, the licensee may be unwilling to pay $1,000,000 in royalties if it expects to only profit $500,000 before paying such royalties. This is a difficult situation, and in a hypothetical negotiation scenario as part of patent infringement litigation it is not uncommon. In the hypothetical negotiation, though, there has to be a resolution. The parties cannot back out.

Now that we have a starting point, which is usually expressed as a portion or revenue, a dollar amount per unit or a lump-sum amount, we can start to hypothesize about where the two parties would have negotiated from there. From here, the analyst will determine the impact, whether up or down, each of the following factors would have on a reasonable royalty:

1. The royalties received by the patent holder for the licensing of the patent in suit, if any;
2. The royalties paid by the licensee for the use of other patents comparable to the patent in suit;
3. Whether the license will be exclusive or non-exclusive, or restricted or non-restricted in terms of territory or with respect to whom the manufactured product may be sold;
4. The licensor's policy of maintaining its patent monopoly;
5. The commercial relationship between the licensor and licensee, such as whether they are competitors or whether they are inventor and promoter;
6. The effect of selling the patented specialty in promoting sales of other products of the licensee;
7. The duration of the patent and the term of the license;
8. The established profitability of the product made under the patent, its commercial success, and its popularity;
9. The utility and advantages of the patent property over the old modes or devices, if any;
10. The nature of the patented invention and the benefits to those who have used the invention;
11. The extent to which the infringer has made use of the invention, and any evidence probative of the value of that use;
12. The portion of the profit or of the selling price that may be customary in the particular business or in a comparable business to allow for the use of the invention or analogous inventions;

13. The portion of the realizable profit that should be credited to the invention as distinguished from non-patented elements, the manufacturing process, business risks, or significant features or improvements added by the infringer; and,
14. The opinion testimony of qualified experts.

The conclusion is ultimately going to be a matter of opinion as none of these factors' impact can be objectively measured. It is not a matter of saying, "The licensor has historically maintained a patent monopoly by not licensing to others. Add 1% to the royalty rate." Rather, the conclusion is going to be based on the entire body of information that is available to you and whether that information was known or could have reasonable been known or estimated at the time of a hypothetical negotiation, the respective bargaining positions of the parties and the strengths of the factors affecting those positions.

A FEW MORE QUICK POINTS

UNJUST ENRICHMENT

In the event that you are faced with the infringement of a design patent, as opposed to a method patent or a utility patent, you will have the opportunity to consider the concept of unjust enrichment. Just as the name says, the implication is that the infringer has unjustly enriched itself. And that unjust enrichment might be substantial.

Maybe it is more than the patent owner's own lost profits or more than a reasonable royalty would provide. In any event, the courts in this type of case will allow the infringed party to recover from the infringing party all the profits made as a result of the infringement. There are certainly complications associated with this; most notably, the impact of the entire market value rule. That is, what did the design contribute to the profits derived from the sale of infringing products?

PUNITIVE DAMAGES

You may be asking yourself, "Why would I ever negotiate a license if all I am going to have to pay in the end is a reasonable royalty plus some nominal rate of interest?" I am not qualified to get into the constituent parts of willful intent, but the court has the ability to apply, among other remedies, treble damages in the event of egregious infringement. If applied, you will not just be paying a reasonable royalty. You will be paying 3 times a reasonable royalty.

PRESENT VALUE OF FUTURE DAMAGES

To the extent that damages are comprised of lost profits related to future periods, those lost profits will need to be discounted back to a date of damages. That date is typically the date of the first infringing sale.

PRE-JUDGMENT INTEREST

To the extent that damages are related to periods in the past, the court usually allows for pre-judgment interest. Effectively, the damages award is treated as though the infringer put the award into an interest bearing account at the time it would have otherwise been earned by you, whether it was a reasonable royalty or a lost profit. The rates used are often court prescribed and are calculated using simple interest, not compounded interest.

CHAPTER 10

Developing and Evaluating Intellectual Property Assets

Leo J. Jennings

Leo J. Jennings is Counsel at Kramer & Amado, P.C., where he practices IP law. Mr. Jennings also developed and teaches as an Adjunct Professor an annual course on "Innovation and Intellectual Property Management" for the MBA program at the Georgetown University McDonough School of Business. He also teaches an annual course on "Global IP Management" at the Franklin Pierce Law Center Summer Institute (now University of New Hampshire Law Center). Mr. Jennings has provided guest lectures on various IP topics at other law schools and the U.S. Patent and Trademark Office Global IP Academy. Mr. Jennings can be contacted at ljennings@krameramado.com.

Innovation and intellectual property (IP) drive the success of many organizations in today's competitive landscape. All organizations, whether corporate, non-profit, university, research or governmental, can benefit from managing the innovation process for protection of the most valuable developments that arise. This chapter addresses ways to promote innovation and decide on IP protection based on a cost-benefit assessment of pursing IP protection. Innovation can include not just product engineering, but also branding, distribution of products, pricing, marketing, and brand development and even overall business plans. This chapter will focus on examples using a technology-driven new product development cycle.

Managing IP is best done using a cross-functional approach, involving internal experts with legal, financial, marketing, engineering, information technology and even human resources backgrounds and perspectives. Keys steps include promoting innovation, harvesting possible candidates for IP protection, evaluating the candidates, and implementing protection activities.

This chapter is divided into two parts. The first part discusses an overall plan for promoting and harvesting innovations, evaluating those innovations, and deciding the suitable level of IP protection to pursue. The second part provides details of how to perform a monetary valuation of the possible IP rights, which is useful during the evaluation phase.

PART I: PROMOTING, HARVESTING AND PROTECTING INNOVATION

An organization will face many individual decision points related to IP. A key to successfully navigating these decisions is to have an overall strategy and a plan, so that the decisions can be made efficiently and with the most positive impact. Some qualities of an effective IP management plan are outlined below.

BUILDING AN EFFECTIVE CULTURE FOR ENCOURAGING INNOVATION

Innovation and invention can arise spontaneously; however, more often they result from a culture that provides resources and structure to drive the process. For example, a company may benefit from setting aside some budget for creative research and development (R&D) that might not lead to immediate profit, thus freeing engineers to push boundaries or explore new areas without undue pressure for quarterly earnings. Other proven programs for encouraging innovation and development of IP protection include inventor awards. Such awards can be a fixed fee paid to each inventor at the time of filing a patent application and/or a fee paid at the time of issuance of the resulting patent. One risk of a flat payment to each named inventor is that it may lead to the naming of managers or other individuals who technically do not qualify as having contributed to the invention. This can be managed by capping the total payout to be shared, or by employing a disinterested party such as an attorney determine who qualifies as true inventors. Real world examples are often approximately $1,000 to $2,000 for each inventor, most often upon the filing of the application, which encourages inventors to be helpful during the application writing process. A smaller amount might be paid beforehand for a submission that makes it past an earlier approval stage, such as a manager approval. The goal is usually to have the amount large enough to have some meaning to the inventor, but not so large as to cause any internal divergence from overall company business goals. Besides monetary awards, many R&D personnel and their managers will be motivated by recognition for their patent activities, such as through company meetings, or having patent activity be included in job performance appraisals. Some companies will hold a periodic innovation summit at which representatives from various departments meet, review and plan the organization's IP strategy.

Other common tactics for promoting IP protection are benchmarking compared to competitors in the relevant field or industry, and setting of numeric goals for new filings such as patent applications. However, numeric counts of patents or other IP assets are inherently rough measures, since they do not take into account the breadth of protection of each asset, nor the commercial importance of each asset. Similarly, when benchmarking it is necessary to keep in mind that a small portfolio on key technology can far exceed the value of numerous patents on minor improvements. This presents a dilemma in that setting of hard numeric targets can drive activity levels upward, but may also encourage pursuit of relatively less valuable cases for the sake of meeting the numeric goal. Also, there is a tendency to rush to meet the activity target at the end of the measurement period such as year-end, leading to hasty and uninformed decisions to file and associated costs. A best approach if using numeric targets is to monitor activity regularly during the year and strive for an even level of filings during the period, unless issues like a large new product launch or a trade show drive a need for more of a batch type process.

A next step is to provide a vehicle by which innovations can be captured for further evaluation. In R&D situations, the tools often used are "inventor's notebooks" and "invention disclosure forms." Inventors' notebooks can facilitate the inventor's creation of a written description of the invention and the date or timeline of creation. An invention disclosure form is a single document, sometimes with attachments, that can be circulated to others for evaluation of the suitability of IP protection. Both documents can have a secondary benefit of proving dates related to the invention in contested legal situations.

Although most persons familiar with patents have a shared concept of the goals and purpose of an invention disclosure form, there is no universal standard for the content or arrangement of one. The form might be fillable electronically, with attachments such as drawings or slide presentation files attached or pasted in, or it might be a paper form filled in by hand with similar attachments. An Internet search will uncover a number of published examples of invention disclosure forms. Before adopting such a form, the IP manager of a company should consider which items are most applicable, and some risks associate with some other items. A few of the most typical items on such forms are discussed below:

- **Inventor names, job titles and addresses.** This should be tailored so the inventors can be easily identified for any company award payments by the human resources department. Also the U.S. Patent and Trademark Office (USPTO) requires a residence address and putting it on the form saves time by having it readily available for the needed forms. The inventors themselves may not have the legal sophistication to determine who actually qualifies legally as an inventor. Due to prestige of the award, sometimes a manager will add themselves as an inventor on every form, and so it is desirable to have the inventorship confirmed at a relatively early stage in the approval process.
- **Title of invention.** This provides a helpful name to refer to the invention that is being documented and/or evaluated for a patent project. However, the actual title of the patent application will likely differ, for legal reasons, from the one originally given by the inventors. This difference can cause confusion down

the road, and so it is often a good practice to assign a "reference number" to each disclosure document, and use that reference number throughout the process. Many organizations include the year of the document as part of the reference number, which is helpful later when scanning various document listings. Also they may include some initials to identify a business unit or geographic location where the disclosure originated or will be handled (e.g. 2011-0123456 / US).

- **Past timeline.** This field provides answers to prompted questions such as when the idea was first conceived, when a prototype was built, and were there any contacts with outsiders. The conception and prototype dates are evidence in the U.S. of dates that may help the company obtain a patent even though another entity files first. The contacts with outsiders helps ensure that the application is timely filed and that the question of joint invention with the outsiders is addressed up front.
- **Background, or prior art.** This is a tricky section. On one hand, the USPTO requires that the applicant identify all "prior art" information that is known to applicant. The standard for knowing what prior information qualifies is a sophisticated legal standard. One goal of this section is to comply with this duty, and typically a simple listing by the inventor of prior patents or journal articles or competitive products can help meet this goal. But if the inventors over-identify prior art, such as the results of a broad Internet search, they can create difficulties for the patent attorney in sorting out what should be submitted. Also in this section the inventors might mischaracterize what is really in the prior art, and create possible admissions that would lead to invalidity of their patent. For these reasons it is legally desirable that this section be brief and narrow in scope, and that the inventors have a detailed in-person discussion with the attorney who will prepare the application. Some inventors lose sight of the fact

that patents are granted on what is new, not what is old, and will provide lengthy expositions of background, perhaps to impress the managers of the value of their project or their knowledge. However experience usually shows that a concise background is more beneficial from the patent standpoint, as long as an attorney interview occurs to ensure that all legally needed prior art is indeed cited to the UPSTO. Thus, it is wise to seek attorney advice on this topic and develop best practices tailored to the industry and company that are shared with inventors.

- **Benefits of the invention.** This section meets competing objectives. On one hand it helps the evaluator make decisions whether to pursue filing, and helps everyone understand why the invention might be patentable. On the other hand, it might also be found to contain admissions that could limit the interpretation of the patent later. It is wise to seek attorney advice relevant to this topic and share guidelines with inventors.
- **Value of, or commercial plans for, the invention.** This section helps managers make cost-benefit filing decisions. However, again this information could be found to create harmful admissions later during patent enforcement, and so might best be handled in a separate document or other communication vehicle. Many large companies assign internal importance "ratings" to disclosures; however, these ratings are assigned by legal staff after the disclosure is evaluated, and are carefully managed with their meanings being qualified to avoid mis-interpretation.
- **Upcoming (or past) dates of disclosure, product release, trade show, publication, sale or offer to sell.** The primary reason for this section is to ensure that a patent application is filed either before any such date or, if only U.S. rights are sought, filed within one year of such date. A consideration here is that these dates

are often fluid and may change after the document is prepared. Also, inventors may not be the most knowledgeable of such dates—in many instances the marketing or sales department pursues these activities ahead of the time of which R&D staff or inventors are aware.

- **Description of preferred embodiments.** This is usually the longest and also most straightforward section from an inventor standpoint. Some inventors are naturally more concise and others more prolific, but this section gives the manager a starting point for a filing decision, and gives the attorney a starting point to interview the inventor(s) and prepare an application. Diagrams are usually attached or pasted in and a technical discussion provided.
- **Signature and date.** This may help the document as evidence to show an invention date in the U.S. The process of witnessing and/or notarization is sometimes added to strengthen this effect. The signature and date lines also help identify the "final version" of the document and that all inventors have read and approved it.

Once the invention disclosure form is created, the process of sorting and evaluating the innovation begins, with the goal of determining whether to pursue IP protection and how much to invest. The size of the organization often determines the decision maker. For example, in a small start-up company, typically one individual is more expert at IP issues and will make the decision more or less independently. This might be the company president, or and individual with IP experience from previous companies, or it might be the company's outside legal counsel providing advice upon which the company leaders will rely. In the end, it is a business decision with inherently uncertain consequences, due to the risks and unpredictability inherent in governmental approval and litigation, not to mention the many unknowns of the business marketplace.

As a company grows larger, it becomes more possible to involve additional individuals in the decision process. A large multinational corporation might have one or more IP committees that meet on a regular periodic schedule to evaluate internal invention disclosures for further action. Various committees may specialize in technology areas or geographic areas.

International issues in patent protection can be very complex and usually require specialized experience to manage. Although patent procedures and standards are relatively well harmonized in the developed world, there are still key differences in style, format and language of applications, and the legal costs in each country often multiply by country or region, so that the managers of an international portfolio can incur higher costs than are often expected by an inexperienced manager. For example, the simple cost of translations can be quite high, and in some countries may be thousands of dollars per patent application during the life of the examination process. Also, patent law is territorial with each nation granting its own patents. Although there is an international patent application process via the Patent Cooperation Treaty (PCT), and there are some regional patent application processes, the rule remains that to obtain patents one must generally take some action and pay some fees in each country where protection is obtained.

The U.S. charges maintenance fees after the U.S. patent is granted, which are due at 4, 8 and 12 years from grant. Most other countries charge periodic or even annual fees (called maintenance fees or annuities) during the pendency of the application and the period in force for each patent. Moreover some countries do not permit as many claims in a single patent as the U.S. does, which means multiple filings could be needed to cover the same scope as one U.S. patent. The costs for multi-country protection can be very high.

As a rough simple example, the lifetime cost of acquiring and maintaining a single patent in the U.S. might be $20,000. The same protection in Canada and Australia, which speak English and have relatively U.S.-similar systems, might be $35,000. The same protection in the U.S., all of Europe, Korea, Japan, India and China might cost $200,000. Protection in the middle east, Russia, South America, more of Asia, and the Nordic countries could push a portfolio over $300,000. A general rule once the number of countries gets to a dozen or more is to estimate the U.S. cost and multiply by the number of countries. However some countries are much more economical than others. Thus, a tradeoff can be made to prioritize spending—perhaps it is better to focus on strong protection in a few nations where enforcement is strong and/or licensing activities can be supported, as opposed to consuming all resources with geographic spread.

Enforcement of patents is not automatic. Ideally a patent owner plans for its maintenance costs and enforcement budget while making decisions on filing for protection. Once granted, the owner still must take legal action to stop infringers or obtain payments or damages. For this reason some business persons refer to the patent as a "right to sue." The costs for enforcement range greatly but can be quite high. A single U.S. patent infringement suit taken to trial can often entail $500,000 to over $5,000,000 in legal fees for the patent owner alone. The accused infringer can face similar legal fees. Due to the high cost, many U.S. patent suits settle during the process. Also in the U.S., contingency lawyers will sometimes take on promising cases. Patent enforcement is notably less expensive in most other countries. However, there is some feeling that the remedies for the patent owner can be weaker in terms of money awarded and/or exclusion effect compared to the U.S.

The decision whether to pursue protection is in essence a cost-benefit analysis, based on the cost of protection, maintenance and enforcement, compared to the value of having the IP right available. The cost of IP protection can vary greatly, but is generally predictable within a certain range. The benefit of the IP right may be more difficult to quantify, but as explained below there are accepted methods to value future IP rights. On the cost side, the amounts will be primarily legal fees, either the use of legal resources inside the company or fees from outside counsel or a combination, and also the long term government maintenance or annuity fees. Other internal costs may be the occupation of time of inventors or business managers. As explained above, a single U.S. patent might entail an eventual outlay of tens of thousands of dollars, and global filing hundreds of thousands. Trademarks are significantly less expensive, and copyrights even less so. Fortunately, if the asset is found to be hard to protect, or is found not valuable in the market, then IP protection activities can be halted, saving further expenses. In fact, the process of abandoning currently pending or granted IP rights in order to save further costs can be one of the most productive decisions to make, since then funds that were formerly going to low-value or weak IP rights can be saved or transferred to more productive IP projects.

Patent research provides a tool that can be helpful in the cost-benefit analysis. A patentability search will often uncover prior patents and other documents that will likely block or limit the scope of available protection if a patent application were to be filed. Depending on the results, it may be best to avoid filing on some inventions if the patentability search shows a low likelihood of success. A clearance search, also called a freedom-to-operate or right-to-use search, can uncover patents of others that would block commercialization of the invention. Advance knowledge of these blocking patents can help the company avoid infringing on the rights of others. Also both types of patent research can often spur the company's development team to think of new ideas. One common instance is where a blocking patent is found and the company develops a new system or method that avoids infringing the blocking patent. This situation of a "design-around" not only helps avoid liability for the company, but also sometimes triggers new approaches to a problem that themselves may be patentable, valuable and even revolutionary.

PROTECTING NEW INNOVATIONS

Once the decision-maker has suitable information, the next step is choosing the appropriate path for IP protection. To determine the level of protection to afford to the innovation, the value of protection should be established. The protection, if obtained, may provide strategic value, such as by blocking competitors, may provide a defensive bargaining chip if an IP battle arises, or may even provide a marketing benefit via advertising of the patent rights to show technical superiority to the customer. In addition to strategic value, the monetary value of the IP right should be considered, which can be determined using methods discussed later in this chapter.

Another question to ask is whether the innovation would have a high value to others. Most IP rights are transferable via sales or license, and in some countries an innovator can obtain significant revenue even without the innovator practicing a patent itself. Again, the monetary valuation methods are useful in this context.

A blend of considerations of the internal and external value, and the monetary and strategic value, informs the protection decision. An invention having high ratings on one or more categories would deserve the highest level of protection, while a lack of monetary value to anyone might suggest de-emphasizing the level of protection sought.

Once the level of protection is identified for an innovation, an organization must then choose the timing and level of investment for IP protection. Patent filings in particular are subject to loss of some or all patent rights if a proper filing is not made with a Patent Office before certain time limits pass. In some cases a filing is needed before any disclosure, sale or publication of the invention. Other actions can also trigger due dates or cause a loss of rights.

Another step in determining which route to take for IP protection is to determine the type of innovation that is present, e.g., trade secret, original work of authorship, distinctive indicator of origin, inventive process, machine, manufacture, and composition of matter. This chapter focuses on patent rights; however a key related right is trade secret protection. A decision to file for a patent will likely lead to eventual publication of the invention, and thus trade secret rights will be lost, which is a cost of the patent protection. In addition the patent will eventually expire, while a trade secret in theory can be preserved indefinitely.

Depending on the value determined above, the level of protection pursued can be adjusted. Patent rights can either be aggressively prosecuted (high protection), or a more low-key approach can be adopted (low protection). A broader patent (possibly more valuable but also possibly more costly to obtain) is more likely in the former case, and a narrower one in the latter case.

MANAGING AN EXISTING IP PORTFOLIO

After an IP right has been acquired, an organization should map out a maintenance plan. Whether it be a trade secret, a copyright, a trademark, or a patent, the questions that were asked while determining the type and level of protection above are still pertinent.

If the innovation is valuable to an organization, it should be maintained and enforced. A patent that is highly valuable to an organization can also be expanded by acquiring IP rights in other countries. Additionally, improvements to the existing innovations can be further protected. On the other hand, if the innovation is not valuable to an organization, but is valuable to others, it can still be maintained and preserved. However, if the innovation is neither valuable to an organization, nor to others, it should be abandoned to avoid the superfluous cost of maintaining it.

PART II: MONETARY VALUATION OF IP

Intellectual property rights can be evaluated as what an IP asset or portfolio is worth in monetary terms. There are multiple reasons why a firm may conduct such IP valuation, some of which are: to manage an IP portfolio; to acquire IP rights; to calculate prospective litigation damages; and to satisfy financial/tax reporting requirements. The preliminary questions that need to be answered when performing IP valuation are: "what is the asset or portfolio being valued (which can include patents and trade secrets, as well as other types of IP rights)?", and "who owns the asset?" The answers to these questions can aid in the selection of a valuation methodology. Over the years, multiple methods have been devised to evaluate an IP asset. They can be classified into one of the following four groups: Cost approach, Income approach, Market approach, and Hybrid methods.

COST APPROACH

This is one of the most fundamental methods for evaluating a "hard" or tangible asset, yet is generally not very informative for IP rights, which are intangible assets. The cost approach can be applied to IP rights, although it is not very useful for making pro-active business decisions about IP. In this approach, the IP owner adds up the investment made in acquiring the asset (which is essentially a sunk cost). Depending on the availability of information, these costs may include research costs, equipment, labor, supplies, marketing, and overhead. Although the value of the asset is relatively easy to calculate based on this method, the asset is often undervalued. Additionally, the cost is generally not strongly correlated to what another entity might pay for the asset (that is, the cost does not take into account the fair market value of the asset). Furthermore, this approach also suffers from a lack of consideration of future income streams, market conditions, and the useful life of the asset. Thus, although the cost approach is a useful tool when assessing the cost of an IP asset from a historical standpoint, it is typically not that informational later on.

INCOME APPROACH (DISCOUNTED CASH FLOW ANALYSIS)

The income approach computes the value of an IP asset by determining the amount of revenue that can be generated from the asset. This approach is also referred to as the net present value approach. A decision tree is used to determine the value based on the following factors:

Total future income attributable to asset

This step requires the user to determine two factors: the sales and profitability of products that incorporate the intellectual property; and the portion of this income that is attributable to the IP asset.

Timing and duration of income stream

The value of an IP asset can vary greatly depending on the time when it is launched in the market. If an asset is launched when the demand for it is high, it can result in great income potential. The timing of an asset depends on multiple factors, including time taken to design, develop, and marketing of the product. Once the product is launched, the duration of the income stream is further influenced by a variety of factors. One of the following approaches can be used to determine the length of useful life of the asset, such as the statutory or legal life (duration of legal protection), functional or technological life (length of time before the technology becomes obsolete), market life (length of time for which there is a market demand for the product), and economic life (length of time for which the asset is producing an adequate return on investment).

DISCOUNT RATE

Once the expected annual cash flow has been forecasted, it must be discounted to its present value. Discounting is a useful tool because not only does it account for the time value of money, it also accounts for some future unknown risk.

Although the income approach is highly analytic, and attractive to investors, purchasers, and licensees, it is difficult to apply in most domains because there is no predicable income stream (the assets are unique and not yet established in the market). Additionally, it is often difficult to unbundle IP assets from the other components of a larger asset (that is, separate the income generated from the IP asset alone, and not the other integrated components). Furthermore, predicting interest rates is uncertain. This problem could be alleviated by using some interest rate forecasting algorithm such as spot rates and forward curves.

MARKET APPROACH

The market approach values an IP asset based on similar assets that have been traded in the recent past. This is a well established approach to valuation. However, most often such comparable transactions are confidential, and the assets are not identical. Thus, determining the value of an asset based on "similar" tranactions is often difficult, if not impossible.

HYBRID APPROACH

The hybrid approach to IP asset valuation simulates a hypothetical negotiation between two willing parties interested in transacting in the asset. One area of development of hybrid approaches has been the use of expert witnesses who testify in patent litigation disputes to advise a judge or jury as to damages that are due to be paid to the patent owner. As a result, there is a community of experience valuation professionals, many of whom have experience as such testifying experts. Some accounting firms specialize in such valuation services and can provide valuation reports for pre-litigation decisons and/or potential investment or corporate acquisitions. Over the years, multiple hybrid approaches have been developed to compute the value of an IP asset. Some of them are:

RULES OF THUMB

This approach utilizes established royalty rates and assumed profit margins to determine the transaction costs and value of the IP assets. For example one benchmark is the 5% rule, where an amount of 5% of a sales amount is assumed as the value of the IP right. Another benchmark used is the 25% rule wherein the royalty for the asset is calculated as 25% of the gross profit, before taxes, from the enterprise operations in which the IP product is used. A primary benefit of these rules is that they are quick and easy to apply, using very limited data. However, the definition of the relevant sales amount can be difficult to precisely measure in view of discounts or price changes, and the relevant profit amount also can vary depending on how various overhead or other costs are allocated against the sales price.

BLACK-SCHOLES OPTION PRICING/REAL OPTIONS THEORY

There are existing IP valuation option-based methods derived from the Black-Scholes options pricing model. Many IP valuation service providers have proprietary options-based IP valuation methods as well. These methods can be very computationally complex. An Internet search can reveal the current status of this theory.

TRADED ON FINANCIAL MARKETS

Some efforts have been made to link IP portfolios to financial market trading. For example, bonds have been issued that are secured by revenues from recorded music portfolios. Public patent auctions have been conducted. However, to date the securitization of IP rights remains a limited activity. As the public and businesses become more aware of the nature and value of IP rights, we can expect a trend toward more securitization.

PATENT COUNTING

Evaluation of a portfolio, as opposed to a single asset, provides temptation to use a counting method, in which a portfolio of one hundred patents appears much more valuable than a group of ten. While this has some intuitive appeal, it ignores that many patents are not of high value, while a fundamental patent can cover large parts of an industry.

MONTE CARLO SIMULATION

This is a refinement of the income approach discussed above. First, a range of values is assigned to variables used in calculating Net Present Value (NPV). Some of these variables could be: price variables such as price premium, additional unit sales; and cost variables such as Cost Of Goods Sold (COGS), and Selling, General and Administrative expenses (SG&A). Once the variables are selected, the next step is to assign a probability to individual values within a range. One of the following probability distribution methods can be used: uniform, triangular, normal, and log-normal. Then, the NPV is repeated 500-1,000 times based upon random selection of probability. Weighted values are assigned to each variable, and multiple NPVs are then plotted by frequency of occurrence, indicating most likely NPV.

SELECTING THE METHOD

A chart (Table 1) is provided to summarize some of the methods discussed above. Testing these methods on real-life situations will reveal that in a given case the different methods can produce startlingly different results as a final numeric value. Yet each method is "accepted." When trying to determine a "true" asset value for internal business decision, it is often best to do calculations using several methods, then compare the results to get some boundary values and begin to bracket in what might be a realistic value based on which method seems to produce the most suitable data. When involved in a negotiation with a third party, it might be best to internally evaluate all methods, but select one for presentation in the negotiation that is favorable to your side.

Table 1: A Summary of Some Valuation Methods

Method	Salient Features	Strengths	Weaknesses
Cost Approach	Add all the investment made in acquiring the asset	• Easy to calculate • Useful in new markets where other approaches are difficult to use	• Undervalues asset • Not strongly related to Fair Market Value (FMV) of asset • Lack of consideration for: · Future income streams · Market conditions · Useful life of asset
Income Approach	Compute the amount of revenue that the asset can generate	• Highly analytical • Attractive to investors, purchasers, and licensees	• Difficult to apply because no predicable income stream • Difficult to unbundle IP assets • Predicting discount rate is uncertain
Market Approach	Values based on similar assets that have been traded in the recent past	• Well established	• Most often such comparable deals are secret • Assets being compared are unique and not identical
25% Rule	Licensor gets 25%, and licensee gets 75% of the profits of the asset	• Ballpark estimate	• Low level of precision • Does not consider true profitability and return on investment (ROI)

CONCLUSION

The processes of encouraging, identifying, valuing and protecting IP rights such as patents is best treated as a continuous cycle. Understanding and applying the best practices in each step can lead to effective management of IP to achieve an organization's objectives.

ACKNOWLEDGEMENT

The author wishes to acknowledge the work of Sumedha Ahuja, law clerk at Kramer & Amado, P.C., for her contributions to this chapter.

Chapter 11

Patent Sales

David O'Steen

David O'Steen is Founder and President of Atlantic Intellectual Property Consulting LLC. Atlantic IP is an intellectual property boutique and patent brokerage specializing in aiding patent holders in realizing the value of their assets. Before helping patent holders monetize their assets, he worked as a patent examiner at the U. S. Patent and Trademark Office. Mr. O'Steen is a registered patent agent and holds a BA in Philosophy, a BS in Engineering (both from Swarthmore College), and an MS in Electrical and Computer Engineering from the University of Florida. Mr. O'Steen can be contacted at dosteen@atlanticip.com.

While strategically patenting key features of products is critical to the success of businesses in many fields, a comprehensive patent portfolio contributes value to an operating company in and of itself. The three common ways of directly realizing value from patent portfolios are litigation, licensing, and sale. Litigation is the highest profile method of generating revenue from a patent portfolio. In this case, the patent holder sues an operating company for infringement of one or more of its patents. If successful, the patent holder collects revenue from the operating company, calculated as a reasonable royalty and/or the profits lost by the patent holder, to compensate for the infringement. In a licensing campaign, the patent holder contacts operating companies and inquires whether they would like to purchase a license. Licensing campaigns can take many forms. In general, the original patent holder retains ownership of the licensed patent portfolio and the operating companies pay a royalty for a nonexclusive license to make, use, or sell goods and services embodying features claimed in the patents.

The final method for realizing value from a patent portfolio is sale of the portfolio. This approach involves the original patent holder placing the patent portfolio up for sale and then transferring title to the assets to the buyer at closing.

All three of these approaches have their strengths and weaknesses. Litigation often brings the highest level of revenue to the patent holder, but it is also frequently the most costly, since it requires the patent holder to hire a law firm, along with expert witnesses, which can push the total cost to millions of dollars. Not all litigations are successful, though. If the patent holder loses a case, it may impact the patent holder's ability to sue other potential infringers or license the patent later on. The patent holder also faces the prospect of invalidation of its patents by the court.

A successful licensing campaign can bring in significant revenue streams for a patent holder. Typically, these revenue streams will be smaller than ones secured via litigation so as to provide an incentive to operating companies to acquire a license and avoid a legal dispute. However, licensing also has several potential drawbacks. Fully licensing a patent portfolio may take years. Also, there is always the possibility of a potential licensee not agreeing to the patent holder's terms. In these instances, the patent holder is faced with either forgoing revenue from the potential licensee or taking the potential licensee to court. Furthermore, a patent holder may find itself in court anyway. Aggressive licensing tactics may give the potential licensee a basis for filing a declaratory judgment action. When a potential licensee files a declaratory judgment in an issue involving patents, the potential licensee typically asks a court to rule on whether its product infringes claims in the patent portfolio and whether these claims are valid. The patent holder is then required to defend its patents in court and also make the case for infringement. The possibility that a licensing campaign can turn into a litigation campaign is a concern for the patent holder. Furthermore, it is becoming increasingly common for parties interested in licensing to start their campaign by filing suit against their target prospective licensees. In many ways, the risk and reward profiles of licensing and litigation are converging.

The sale of patent assets is the most direct approach for revenue realization. The timeline from start to finish is generally the shortest of the three. Moreover, there is a decreased chance of unintentionally ending up in court. Outright sale, however, normally results in lower compensation to the original patent holder than either litigation or licensing.

While the remainder of this chapter will focus exclusively on the patent sale process, it is important for the patent holder to keep all options and their risk and reward profile in mind when making decisions regarding its patent portfolios. From this point on, the terms "patent portfolio" and "intellectual property" will be used interchangeably to refer to any group of patents and applications held by a company or individual.

WHY SELL?

Patent sales can be an excellent option for companies looking to better leverage their assets. For companies looking for capital for growth, sale of patent assets can be a smart alternative to either borrowing money or selling equity in the firm. Patent sales do not require the company to find collateral or make interest payments. Furthermore, the present ownership does not have to dilute their interest in the firm. Patent sales are also an option for companies or investors exiting certain markets. As operations wind down, sale of patent assets may partially offset losses accrued by winding down operations. This last option may be of particular interest to venture capitalists looking to recoup investments from failed companies.

WHERE TO LOOK FOR INTELLECTUAL PROPERTY TO SELL

The selection of intellectual property for sale is one of the most difficult aspects of the entire process. An excellent starting point for most companies is to look at intellectual property in areas no longer of active interest to the firm. Often, after mergers and acquisitions, there is intellectual property owned by the combined firm that is no longer being actively exploited by the firm's operations. Also, companies may exit certain markets or abandon certain product lines, but nonetheless have patents covering features in those fields. In such cases, it is important to separate the potential value of the patents in these fields from the business decisions that led to moving away from operations in these fields. Before abandoning these patents and essentially forfeiting any future opportunities, it makes sense to investigate whether they would be of interest to any potential buyers. Of course, companies can also sell patents from areas where they are still operating. Depending on their market and financial position, this may be an appealing option. As will be discussed later, there are ways to structure the sale of intellectual property that allows the company to profit from the sale, but still retain the right to practice their inventions after the sale.

Patent sales are also an option for bankruptcy trustees and estates. A bankruptcy trustee may be able to strip the patents from the remainder of the assets and sell them without affecting the valuations of other assets managed by the trust. A patent sale is an appealing alternative for creditors otherwise resigned to letting these assets go abandoned.

Finally, it is important to make sure that a patent holder holds the title to any assets that he wishes to sell and that he is sole owner of the asset and that there are no liens or claims on the patent assets by others, such as a creditor or an alleged inventor whose rights were not assigned to the patent holder. Also, the patent holder must not have any prior contractual obligations hindering or restricting a sale or transfer of ownership of the assets.

WHAT ASSETS SELL?

The market for intellectual property is dynamic, and as different participants enter and exit the market, factors affecting the sale of intellectual property will change. Throughout these changes, though, there are a few broad trends of which potential sellers should be aware. The first trend addresses which patent jurisdictions garner the most interest from buyers. The mostly highly sought patents are issued in the United States of America. U.S. patents are highly valued for several reasons. A patent grants rights which are geographically limited to the issuing jurisdiction—a U.S. patent protects holders from the unauthorized making, using, selling, and offering to sell by others of products and services embodying inventions claimed in the patents in the U.S. The U.S. is the largest market for enforcing a single patent. This fact will be true until China or another country overtakes the U.S. as the world's largest economy, or until Europe further consolidates its patent system.

The U.S. has another advantage: its legal system. The federal court system has experience in handling complex infringement cases. Judgments are readily enforceable in the U.S. Other administrative venues important to patent holders, such as the U.S. Patent and Trademark Office (USPTO) and the International Trade Commission (ITC), operate in a professional and predictable way. All of these factors together make U.S. patents the most valuable and sought-after patents on the open market.

In some cases, buyers are interested in patents issued in other jurisdictions. In today's market, patents issued in larger European jurisdictions, like France, Germany, and the United Kingdom, as well as Japan, may also attract buyers, albeit at lower prices than a U.S. patent with comparable patent coverage. As legal systems around the world assert their political independence and mature, patents issued in other jurisdictions will garner more interest.

It is not uncommon for patent holders to have applications and patents issued in a variety of jurisdictions. This allows the seller to present multi-jurisdictional coverage of product features to prospective buyers. Certain types of buyers favor this situation and are willing to offer higher sums to the sellers for portfolios with patents in key jurisdictions. In situations where the patents and applications directed to multiple jurisdictions are legally related through priority claims, most buyers will insist that they are offered together in the same sale.

The market is primarily interested in patents issued either in the U.S. or in other major jurisdictions. Typically, for a sale to be successful, a granted U.S. patent is part of the asset list. It is not unheard of, though, for a portfolio to sell that is mostly or entirely comprised of patent applications. Because applications are not yet enforceable and their claims are subject to change, valuation of applications is extremely difficult. Furthermore, the new owner of these applications has to finish prosecuting them. For these reasons, many buyers of intellectual property are uninterested in acquiring application-only portfolios. Others only buy application-only portfolios when they are priced at a deep discount from what an equivalent portfolio of issued patents would sell for.

When evaluating which assets should be put on sale, it is important for patent holders to keep in mind potential use and infringement related to the assets. No buyer is interested in patents claiming inventions that are highly unlikely to be practiced. Buyers, no matter what their strategy or business model, want intellectual property that is contributing, or will contribute, to the commercial or technological success of goods and services. In general, buyers will place a higher valuation on intellectual property that is already being infringed. This fact is often counterintuitive to many patent sellers. Many sellers assume that buyers are only interested in "cutting edge" or "next generation" patents, and that buyers are uninterested in intellectual property already being practiced by companies. Many patent holders believe that buyers primarily acquire intellectual property to develop future product lines. This is generally not the case.

Some buyers are interested in patents that are not presently being practiced. Salable patents that fall into this category are in fields that are rapidly evolving and hold great potential for profits. In more mature fields, dominant products are the rule and alternative approaches have essentially been relegated to the sidelines. In new commercial fields or fields undergoing rapid changes, there often is not a dominant product, or even a dominant business model or product type. Because it is difficult to predict how these dynamic markets will develop, intellectual property buyers will hedge their bets. They are willing to take a chance on patents covering promising technologies. In these instances, patent buyers are more open to acquiring patent portfolios not presently being infringed.

Transactions of intellectual property take place across a multitude of technological fields. In the recent past, though, certain fields have been favored. Fields with the highest volumes of sales tend to be those based on electrical and computer engineering. Specifically, patents in communications, computer memory, search, document and media presentation, and consumer electronics have all sold over the last few years.

Determining which fields will attract interest from buyers in the future is difficult. The fields listed above will continue to retain the interest of buyers because of their commercial success. As other commercial markets develop, intellectual property sales in those fields will also pick up. One broad trend worth noting is that the patent market tends to lag behind the commercial market. In general, patent buyers become active in acquiring patents in technological fields where commercially successful products already exist. As fields like renewable energy and biotech develop commercially, buyers will become increasingly interested in acquiring patents in these fields as well.

PATENT BUYERS

Patent buyers fall into two broad groups: operating companies and non-practicing entities. Operating companies include any company that derives at least some of its revenue from providing goods and services to customers. In general, operating companies active in the patent market are large, multinational corporations. These companies have the resources to evaluate patent portfolios as they become available and are able to pay the purchase price at closing. They also have the most to gain from acquiring intellectual property from other parties. Operating companies generally are defensive buyers of intellectual property. They acquire intellectual property to protect existing and future lines of business. For them, acquiring the patents outright is a cheaper alternative to allowing someone else to control the portfolio and assert it against them. Operating companies are normally content to simply acquire the patent portfolio and do not plan to actively enforce the patent portfolio against others.

The other main group of patent buyers is more heterogeneous. Non-practicing entities comprise three main types of enterprise: Patent Licensing and Enforcement Companies (PLECs), patent aggregators, and private and venture capital investors. PLECs, often referred to by the pejorative term "patent trolls," are the most common and best known. PLECs acquire intellectual property for the purposes of licensing or enforcing it in a court of law. PLECs do not make, use, or sell goods or services embodying the patents that they own. They generate all of their revenue from the royalties due to them from operating companies practicing the claimed features of their patents.

Patent aggregators are the second main group of non-practicing entities. Patent aggregators are organizations that acquire intellectual property to ensure that it does not fall into the hands of PLECs. These organizations do not license or enforce any of the patents that they acquire. These organizations acquire patents that have commercial significance to their investors or clients. Patent aggregators may be structured as non-profits. They may also generate profits for their investors by charging subscription fees.

The last group of non-practicing entities includes private equity and venture capital investors. This group includes investment groups and private buyers that occasionally invest in intellectual property. Depending on the circumstances of the acquisition, these groups may act as defensive buyers to protect existing lines of business, or they may eventually license and enforce their acquired patents. In other cases, they may be looking to reap a capital gain on their investment by reselling their portfolio to others.

For potential sellers of patents, it is an exciting development that so many different types of organizations have taken an interest in acquiring intellectual property. There are benefits for dealing with each kind of company. Operating companies, like the major technology firms, are well funded and willing to pay the entire purchase price in one lump sum at the close of the deal. For patent holders looking to quickly liquidate their assets, this can be a decided advantage. The various non-practicing entities, on the other hand, have a heightened interest in acquiring new portfolios since their sole source of revenue is patent enforcement. This fact makes non-practicing entities quick to respond and react to new portfolios being placed on the market. In certain circumstances, these companies are willing to structure deals which provide the patent seller with revenue years after the close of the deal.

It is important for any potential seller to be aware of who is active in the patent market. By understanding the audience for their assets, patent holders can better understand who the likely purchaser will be and how a valuation will be made. It will help the potential seller navigate the marketing and sale process.

VALUATION

Two basic approaches inform the valuation of patent assets. The first approach uses Net Present Value (NPV) calculations. In a net present value calculation, likely future revenue streams are discounted back to the present day and summed together to determine the value of the asset. This technique is well-known in finance. Net present value exercises are often used by large patent holders to quantify their holdings for accounting, tax, and other purposes. The second basic approach relies on market information to determine the value of the asset. This approach involves looking at comparable transactions and interpolating from those outcomes.

Both of these methods have their drawbacks. Net present value estimates may be inaccurate because it is difficult to populate all the inputs necessary to perform the calculations. The comparable transactions method often provides inaccurate results because there simply are not enough data points in most patent fields from which to draw sound pricing conclusions. Furthermore, the patent market is quite dynamic. As certain companies become active and others become less so, the patent market changes, and with it, the pricing of patent assets also changes. Also, market participants are not insulated from the wider economy. Depending on their own financial outlook, they may be less willing to adhere to past pricing trends. Instead of relying solely on one approach to valuation, it may be more fruitful for a patent seller to keep both in mind when approaching the patent market. Moreover, patent sellers need to be aware of the basic characteristics of their portfolios that affect value, regardless of which approach is used.

The single most important characteristic to consider when deciding the pricing of a patent portfolio is potential use or potential infringement. The presence of infringement in a patent portfolio not only drives its value, but also increases the likelihood of a sale. When there is a reasonable case that a portfolio is being infringed, it is important to identify the patents and their claims that are alleged to be infringed. Once these patents and claims have been identified, it is important to identify products and services that are infringing the claims. Patent buyers draw a huge distinction between vague and general assertions of infringement and specific and supported claims of infringement. Statements of the structure "Company X is infringing this patent" or "Every X of this type of product infringe this patent" are not sufficient to generate interest in the marketplace. Instead it is better to list products and to specifically name the features of the products that embody the claims of the patents. Once the products and the commercial markets directly impacted by the portfolio are identified, the size of the market can be quantified. Documentation of past and present sales is the best way of quantifying the size of the impacted market. Furthermore, it is helpful to have reliable and credible projections about the outlook for the affected commercial markets. Again, having specific market data about the exact market directly impacted by the patent (or patents) for sale is beneficial when making an accurate valuation. For instance, if a patent covers a feature on smart phones, it is better for the patent holder to derive valuation expectations based on smart phone market figures, not on the total market for all cell phones. This approach will give the patent holder a better idea of the value of the portfolio and keep the patent holder's valuations in line with valuations made by prospective buyers.

When the seller has determined that the patent portfolio is not currently being infringed, it still should quantify the markets potentially impacted by the portfolio. For the patent portfolio to have any chance of selling, it must impact a potentially valuable commercial field. This means that the seller should have a clear idea about where infringement could occur. For example, if the patent reads upon future wireless systems, would the patent read on features found in the wireless client, in the base stations, or in both? If the claims in the portfolio are likely be practiced in separate fields, it is important to identify each one. Furthermore, since the patent is not being currently used, it is important for the potential seller to specify how the claims in the patent will be practiced and justify these conclusions. For instance, are the claims of the present patent a feature that may be included in a future device, or are the claims essential to any implementation of a next generation technology? Once again, it is important that the patent holder clearly identify the markets potentially impacted by the portfolio and clearly demarcates and quantifies them. The best approach for accomplishing this is to use reputable research reports, company data, and third party market data.

There are some clear trends which impact patent valuation. The first is that the larger the market in terms of revenue, the better. Larger markets hold the potential for larger revenue streams from patent royalties. The potentially larger royalty payments boost the value of the patent, as would be expected under a net present value analysis. The second trend addresses the patented features' role in the infringing device. In general, patents which include claims that go to the heart of a product or service are worth more than patents that include claims addressing tangential or ancillary features. Non-essential product features can be designed around or eliminated if they are found to infringe patents. The third trend is that buyers pay more for present infringement than for potential infringement in the future. Future infringement is uncertain, and buyers factor in this uncertainty into their offering price.

As has been previously stated, the presence of infringement also affects the likelihood of sale. Moreover, it affects which parties are likely to bid on the patent portfolio. Infringed intellectual property has the highest likelihood of sale. Buying infringed intellectual property fits nicely into the business models of all major types of patent acquirers. Operating companies with goods and services impacted by the portfolio may decide that acquiring the portfolio is the best option available to them, while patent aggregators will acquire the portfolio to protect their clientele. PLECs prefer to buy currently infringed patents because it allows them to immediately transition over to licensing and enforcement of the patents after the deal has been closed. Technologically well-positioned but non-infringed patents have fewer takers. Generally, pure PLECs will not acquire non-infringed intellectual property because it will not generate revenues in the near term. Patent aggregators tend to be most concerned with patents that directly impact their clients, and so focus more on infringed patents. Operating companies may, in certain instances, acquire presently non-infringed intellectual property that they think may be crucial in the future. If a company sees a portfolio on the market that may impact future products, they may acquire it to avoid litigation and royalty payments in the future.

Another major factor when considering the value of a patent portfolio is prior licensing activity. A license is a contract between the patent holder (licensor) and another party (licensee) giving the other party rights to practice the patent holder's inventions. Licensing is a complicated process worthy of its own treatment, but it is important for patent holders to understand the impact of several common licensing arrangements on the value of their patent portfolios. Licenses can first be broken down into two broad categories: exclusive and nonexclusive licenses. An exclusive license grants the licensee the exclusive right to practice the inventions claimed in the patent portfolio. In other words, the licensor is contractually obligated not to license the affected patent portfolio to any other companies. A nonexclusive license is a license where the licensor has the option of licensing the patent portfolio to more than one licensee.

Patent licenses can also be transferrable or non-transferrable. A transferable license is a license whose rights can be transferred to another party by the licensee. A non-transferrable license is one that cannot be transferred to another party. Typically, a non-transferrable license only grants the licensee and its affiliates, clients, and fully-owned subsidiaries use of the licensors' portfolio. Any use by entities outside of these groups is not covered by the terms of the license. Licenses may also be sublicenseable. Sublicenseable licenses allow the licensee to extend the terms of the license to others without giving up any of the privileges of the license. In other words, the licensee can act as a licensor of the patent portfolio and pass along or sell the licenses to other operating companies.

A special term found in many patent licenses is field-of-use. A field-of-use license gives the licensee the right to practice the claimed inventions of the patent portfolio, but only in a certain field. These arrangements can make tremendous sense for both the licensor and the licensee. From the licensee's perspective, the field-of-use license may be as useful as a general license. If the licensee is acquiring the license for a line of products in a single field, then there is no need to hold out or pay more for a general license. Also, the licensor may also extend to a licensee an exclusive field-of-use license. This license allows for more than one licensee, but for any given field, there will only be one licensee. Field-of-use licenses also appeal to licensors.

The licensors can tailor agreements to better fit the needs of certain licensees, potentially raising revenues from the patent portfolio. Licensees generally pay more for exclusivity because it guarantees that their competitors will not be able to secure a comparable license.

The existence of previous licenses can have a both a positive and negative effect on pricing. Exclusive licenses tend to have a severely detrimental effect on valuation. An exclusive license prohibits any current or future patent holder from further licensing the affected patent portfolio. The patent holder may also have to consult with the exclusive licensee about such issues as enforcement. Under these circumstances, few buyers will be interested in acquiring patents that have been exclusively licensed. Transferrable licenses also can prove to be a barrier to sale of a portfolio. Different buyers will express different concerns about transferrable licenses. For an offensive buyer like a PLEC, the fact that a transferrable license exists will undercut its ability to license and enforce the portfolio in the future. Any potential target of the PLEC will have the option of buying a license from the original licensee and thereby gaining protection from the threats of the PLEC. Defensive buyers may also balk at buying a patent portfolio with a transferrable license in existence. The defensive buyer is normally only concerned with acquiring rights to the portfolio that give it freedom to operate (that is, to make, use, or sell products embodying claimed features of the patents). If it is cheaper to simply buy the license from an existing licensee, the defensive buyer will do this instead of acquiring the portfolio. It is generally viewed that patents with licenses that allow for licensees to offer sublicenses are essentially devoid of value. The concerns are similar to the ones above involving transferrable licenses but more profound. Offensive buyers will be concerned because potential licensees will have other ways to protect themselves aside from buying a license from the new patent holder. Defensive buyers will look to secure a better deal through a sublicense from one of the existing licensees.

The effects of field-of-use terms in a licensing contract are more difficult to discern. Fields-of-use limitations that were once cut-and-dry may have blurred, and so it becomes more difficult to determine how these licenses affect value. This issue means that the terms of the field-of-use contract are extremely important in deciding who has a license and a right to practice the inventions and in what manner. For the sake of simplicity, some buyers may treat non-exclusive field-of-use licenses as a non-exclusive license and an exclusive field-of-use license as an exclusive license.

From the discussion above, it would appear that existing licenses would only serve to eliminate value from a portfolio. This is not true. Licenses can help demonstrate the value of the patents. Provided that the licenses do not gut the value of the portfolio (like in the case of a license which grants sub-licensing rights to a licensee) or severely restrict the future owner from exploiting their assets (like an exclusive license of some sort), they can demonstrate value to prospective buyers. This is because of the difficulty of licensing patents. Operating companies do not in-license (license other organizations' patents) casually. If a portfolio placed on sale has been licensed by at least one major operating company, it means that the portfolio received a "vote of confidence" from the company. Also, if terms of the license can be revealed and the royalty rates and terms are favorable to the licensor, then this can be an additional boon to the patent holder. Preexisting revenue streams can help fix a purchase price by demonstrating what kinds of returns are possible. In this respect, royalty rates set by preexisting agreements can cut both ways. If early licensing agreements set the royalty rates lower than what the licensor hoped, then these rates may become fixed in the minds of prospective buyers and used as a basis for determining their bids.

From the perspective of operating companies, preexisting licenses are going to be viewed either neutrally or positively. Most operating companies will focus less on the lost revenues from preexisting licenses since they normally buy for defensive reasons. They may look on the licenses favorably because they show that at least one of their peers thought strongly enough about the technology covered in the patent portfolio to license it themselves.

Non-practicing entities will take a more nuanced view of preexisting licenses. Non-practicing entities acquire intellectual property for its offensive uses and will take into account how much of the potential market for a given technology covered by a patent portfolio has already been licensed. Consider the following example. A patent portfolio claims essential features of products in a certain market, and there are five major companies selling these products in the U.S. The present patent holder has already licensed to two of these companies, which represent roughly half of the market. Non-practicing entities will take all this information into account. On the one hand, the non-practicing entities will view this favorably: two major companies have done extensive due diligence and taken a license to the portfolio. This fact strongly suggests success in future licensing and enforcement and will bolster the sale price of the patent portfolio. On the other hand, they will look at the preexisting licenses as already accounting for half of the market, leaving them with only half the market for new licenses. Since preexisting revenue streams tend to stay with the patent holder, or the royalty payments may have already been made (in the case of the licensee fully paying up at the start of the license), this would exert downward pressure on the sale price of the portfolio. On the balance, though, situations like the one described above are good for the seller. The strength of the patent portfolio has already been demonstrated due to the existing licenses, and the seller can generally expect a favorable outcome from the sales process.

One licensing situation happens quite regularly and demands special attention: cross licensing. Often, operating companies avoid legal action with each other by cross licensing part or all of their portfolios. A cross license is generally an agreement between two companies that give each company a license to large portions or all of the the other's patent portfolio. These licensing agreements are desirable to each operating company because they greatly reduce the chances of litigation occurring between the two companies. The effects on pricing for any portion of the affected portfolio, however, can be negative. Provided that the terms of the cross-licensing agreement stipulate that the license to the patent portfolio is non-transferable, non-sublicenceable, and non-exclusive, then the patent portfolio may still be salable. However, the cross license may still exert a downward force on the sale price because it takes potential targets for licensing and litigation off the table for the new owner and therefore limits potential future revenue for the new owner. Because a cross-license is usually a broad license that covers many or all of the seller's patents, there is no guarantee that any portion of the portfolio could be successfully licensed on its own.

The patent owner should also be aware of other encumbrances that can affect the salability and value of the patents. The first form of encumbrance is participation in any standards body. If the patent holder has previously contributed patents to a standards body, then their use and licensing are at least partially determined by the terms of the standards body. Depending on the terms of the standards body or pool, this may disqualify the patents for sale or severely limit the pool of interested buyers.

Another factor that a potential seller should keep in mind is whether any government funding was used in the development of the patented technology. If so, then the patents are subject to the Bayh-Dole Act. This act may limit who can acquire the patent portfolio and under what conditions. For example, non-practicing entities tend not to acquire patents subject to the Bayh-Dole act because they do not plan on directly commercializing them, one of the stipulations of the act.

Litigation is the clearest test of a patent portfolio's worth and has a major impact on the value of a patent portfolio. In a trial for patent infringement, several outcomes can occur. The first is that the patent holder's patent will be partially or fully invalidated. This is clearly bad for the patent holder. Claims invalidated by the courts cannot be resuscitated by the U.S. Patent and Trademark Office. If a patent has been fully invalidated, it is worthless. If it has been partially invalidated, the patent may still retain value if the remaining claims are broad enough so that commercially important technologies still infringe it. Even if the validity of the patent is upheld, the court may find that the defendant is not infringing. A judgment of non-infringement also hurts the value of the patents at issue. Aside from preventing the patent holder from collecting from the defendant in litigation, it casts potential doubt in the minds of future buyers and licensees whether the patents can be enforced against others. Finally, a court may find that the patents asserted by the patent holder are valid and that the defendant is infringing them. This result can greatly improve the value of the patents at issue. Outcomes like this demonstrate that the patents are valid and can be enforced. Prospective buyers can also reason that successfully enforced patents will be easier to license in the future.

THE SALE PROCESS

Over the past few years, several methods to brokering intellectual property have emerged. Each process has its own advantages and drawbacks. Patent holders, along with their advisors and brokers, should weigh each approach to decide which makes the most business sense. Before discussing seller-initiated brokering (also referred to as sell-side), it is important to discuss buyer-initiated (buy-side) deals. In a buyer-initiated deal, an interested buyer, normally through a broker or other agent, contacts the patent holder and inquires about the assets. If the patent holder is receptive, negotiations move on to sale price and terms. Once agreement is reached, the deal is concluded. In cases such as these, the patent assets never make it to the open market. Only the two parties, along with their consultants and representatives, are engaged throughout the process. The same factors that drive pricing on the open market drive price discussions in this context.

The primary benefit of the buyer-initiated deals is the relative speed with which they can be closed. Because the buyer has already targeted the portfolio for acquisition, time does not have to be spent marketing the portfolio. Typically, the buyer has already started its due diligence, and the whole process can move quickly towards agreement on terms and price, and closing. The tradeoff for the brevity of the process is the lack of alternative buyers. Because there are no competing bids, it can be difficult for the seller to determine if it is getting fair market value for the assets. Furthermore, many patent assets that have sold well on the open market were never part of a buyer-initiated deal. Patent holders who are contacted by a reputable company or agent should consider a buyer-initiated deal, but they should also keep in mind all of their options.

The three major platforms for seller-initiated brokering are the online platform, the auction platform, and the private sale platform. All three approaches can be broken into three broad steps. The first step of a sell-side patent sale is marketing. After the seller and the broker enter into a representation agreement, the broker prepares documentation and marketing materials for prospective buyers. The broker then circulates these materials to prospective buyers. The second step of the process is receiving bids. This step can vary greatly across brokering platforms, but it typically involves the broker receiving bids from all interested buyers and then determining the winning bidder. The final step is finalizing due diligence and closing. Again, the specifics of this step vary across platforms but involve clearing up any outstanding issues with the patents and closing the deal.

The online platform leverages the reach of the Internet to market the assets and achieve a successful sale. The online platform begins with the broker posting the assets on a website. The site may be either open to the public or may be limited to registered buyers. A listing of the assets and an invitation to bid will typically be posted online. There may be a deadline for bid submissions. Prospective buyers can then examine the listing and contact the broker if they are interested in making a bid. If agreement to terms and price can be reached between the two parties, then the deal moves towards close.

The second platform for transacting patents is the multi-lot live patent auction. These auctions tend to use the English auction as their preferred format, and are very similar in style to the fine art and fine wine auctions held by major auction houses. While the terms of the auction may vary a little, there are some general rules. English-style auctions are highest-bidder-wins auctions. Bids start low and end high. While the day of the auction is the most exciting, it is important for sellers to understand the process that leads up to the auction event. Solicitation for lots typically starts months in advance of the auction date. At the beginning of the process, the auctioneer starts evaluating lots from various sellers and judging which ones are good fits for the auction. A lot in a live, multi-lot auction is a patent or set of patents and applications. Lots generally have at least one U.S. patent, but may also include U.S. applications, foreign patents, foreign applications, and abandoned or expired foreign and domestic patents and applications. Before marketing of the patent portfolio begins, the auctioneer and seller negotiate the listing for the assets. Patent auctions differ from other patent transaction platforms because most of the negotiations take place between the auction house and the seller before the final buyer is named or the auction is even held. The principal points of negotiation are the price listings. Depending on the practices of the auction house, there may be one or two key prices affixed to the lot: the reserve price and the estimated price. In an English-style auction, the reserve price is the price that has to be met by the highest (in other words, winning) bid for the lot to automatically transact. In an English-style auction, if the highest bid is below the reserve for the lot, then the transaction does not go through and the seller retains its assets. If the reserve price is met or exceeded, then the seller is contractually obliged to sell its lot to the winning bidder. The purpose of the reserve price is to protect the seller. In establishing a reserve price, the seller should realistically consider what is an acceptable minimum sale price for the portfolio. Setting the reserve too high may eliminate any real chance of success at the auction. Sometimes in an English-style auction, an expected value is also used. Generally, the expected value is a price that is published for each patent lot that gives prospective buyers an idea of how much to bid for the lot. In instances where an expected value is used, the reserve value is typically kept confidential. In auctions where both a reserve and an expected value are used, it is important to make sure there is a reasonable relationship between the two values. If a reserve is set at a low price but the expected value is much higher, bidders may be scared off. They may judge the expected value as too high and not follow through with their diligence or bid on the asset at all. Reserve and expected values should be chosen at a price point that will generate interest from multiple bidders. Gaining and sustaining the interest of multiple buyers is the key to a successful auction for the seller. The dynamics of the auction make it possible for the sale price to rapidly increase on the momentum of multiple bids. Without multiple bids, though, the price of the lot simply stalls.

The marketing for lots in the auction is performed by the auction house. Because the auction is an event, its date is well known to buyers in advance. In the months leading up to the auction, the auction house will release a listing of the lots up for auction. The house may post information online or release a catalog describing the patent lots. Furthermore, they may also prepare materials specific to each lot to circulate to buyers. During this marketing period, prospective buyers conduct their due diligence for all lots that interest them in the auction.

Once a winning bidder has been named for a lot, the transaction generally is finalized shortly thereafter. In the auction format, the transaction will be completed using the auction house's standard assignment and intellectual property sale agreements. These agreements represent "best practices" in intellectual property transactions and contain terms that would be agreeable to practically any buyer. The seller signs the assignment document and the IP sale agreement before marketing for the auction begins. The winning bidder is contractually obligated to pay the auction house for the patent lot within a fixed time period after the date of the auction. The remainder of the process focuses on the seller fixing any issues the buyer may have, such as chain-of-title issues.

The most common process for selling patents is the private sale platform. The private sale platform is when a seller engages a broker to market its patent portfolio to a select list of buyers. Once the decision has been made to take the patent assets to market, the broker starts assembling a marketing package describing the patents for sale. The broker also starts assembling a list of target acquirers for the portfolio. In these two steps, the patent holder may help with both the marketing materials and with the target lists. Once all the materials and target acquirers list has been finalized, the broker starts the marketing process. During this marketing process, interested acquirers start their due diligence. The initial marketing and diligence period is around two to three months. In a private sale, the time frame for the initial diligence period is extremely flexible. Depending on the needs of the seller and the complexity of the patents on offer, the period may be shortened or extended. At the end of the initial diligence period, initial bids from the target companies are due to the broker. The initial bids may be formal bids, or they may merely demonstrate interest on behalf of a prospective buyer. Depending again on the needs of the seller and the complexity of the patent portfolio, there may be further rounds of bidding until a final winning bidder is named. The process then transitions over to finalizing the details and closing the deal. At this point, the principal terms of the deal have been agreed to by both the buyer and the seller. What is left is cleaning up any chain-of-title issues that the buyer may have and signing the IP sale agreement and the assignment document. The IP sale agreement and the assignment document may be provided by either the buyer or the broker. The buyer may also provide to the broker an itemized list of issues they have flagged for resolution by the seller to speed up the time to close.

These three platforms differ from each other in a variety of ways. In a private sale, practically all parameters of the process can be tailored to best fit the seller and the technology involved in the transaction. The type of materials used for marketing can be altered to best fit the situation. Sales materials vary from portfolio to portfolio and from broker to broker, but certain approaches are widely used. First and foremost, the marketing materials tend to include infringement analysis. Most likely, the infringement analysis will be in the form of claim charts. Claim charts are documents that compare the language of a claim in a patent with the features of an allegedly infringing product or service. Furthermore, the marketing materials typically include product segment analysis. This analysis provides prospective buyers with estimates of the size of the markets affected by the portfolio. It also may include estimates on these markets' future growth. Finally, the marketing materials may include financial analysis where reasonable royalty rates for the patents and markets affected are introduced and their effect on royalty streams is discussed. Each part of the marketing materials serves to identify and quantify the value of the portfolio and give prospective buyers an idea of how the portfolios relate to various commercial markets.

The time allowed for the initial marketing and due diligence period can be changed depending on the needs of either the seller or the buyers. In a private sale, there may be multiple rounds of bidding. Each round of bidding may be separated by periods for additional diligence that can be tailored for the specific issues pertaining to the portfolio. By contrast, the auction platform is a highly standard and inflexible process. The date of the auction is set well in advance. The auction house publishes their catalog or materials with all the lots available for sale at a point well in advance of the auction date. These events bookend the initial due diligence and marketing period and the auction date cannot be extended. In an auction, there is only one round of bidding. In the online platform, the initial period for marketing and diligence can be unbounded. Marketing materials used in the online platform may vary in style and substance from the other two platforms.

The types of deals that are available to the seller are affected by the chosen platform. In most online and private sale engagements, the buyers can offer deals with a variety of different structures. This is not true in patent auctions. In patent auctions, only all-cash deals are accepted to allow easy comparison of bids. In a private sale platform, the broker can limit the target list to best suit the desires of the seller. In a patent auction, all registered bidders can bid on any lot available at the auction. Depending on the nature of the online platform, the seller may be able to limit who is able to bid on the assets for sale.

The seller's choice of platform thus should take into account the trade-off between flexibility and quality of service and cost. Typically, online venues tend to offer the lowest cost for their services to their clients, but also the lowest level of service. The types of materials assembled by the online broker may not be as comprehensive as those assembled for the other two platforms. Furthermore, the online operator may not conduct further marketing or solicitation of bids aside from posting the assets online. Online venues may be a good fit for sellers unable to list their assets in a patent auction or engage a broker for private sale of the assets.

The auction platform provides many advantages to sellers. Auction houses widely market their events to the public and prepare marketing materials for the offered patent lots. They also may follow up with certain buyers and market specific lots to these buyers. The auction itself acts as a hard deadline that spurs buyers to undergo their due diligence in a timely manner, which can be an issue in other platforms. Also, by having the seller execute the assignment document and IP sale agreement, the time between the auction event and the close of the deal can be shortened. The principal downsides of the auction platform stem from the rigidity of the process. The auction date cannot be changed. If issues surrounding a patent portfolio turn out to be complex, there is no way to give prospective buyers more time to conduct their diligence. In the auction process, anyone who registers for the auction is free to bid on any lot. This means that seller has no way of excluding parties, like competitors, from the sale. Also, auctions publicly advertise assets for sale and their results are also made public. For many sellers, the publicity is an added bonus. Some sellers, however, may not be interested in making their transaction decisions or results public. For them, the patent auction may not be a wise choice. Furthermore, the very structure of bidding constricts the types of bids a seller can accept. All bids at an auction are all cash, lump sum bids. Sellers receptive to bids with other structures may find the auction platform limiting. Commissions and fees are another consideration. For patent auctions, they tend to lie somewhere in between fees for online venues and private sale engagements. The auction house may charge a listing fee for an auction lot and charge commissions based on the sale price of the lot.

The private sale platform's advantages stem mostly from its adaptability to fit the needs of any seller or portfolio. In a private sale, certain prospective buyers can be targeted and others can be excluded from the process. Time periods for due diligence periods can be altered depending on the requirements of the parties. The types of materials assembled for marketing and aiding in due diligence can be tailored to fit the specific needs of the seller and portfolio. There may be multiple rounds of bidding. The broker and the seller have the option of accepting bids with a variety of structures. This last advantage should not be understated: many buyers may not be willing to pay a lump sum for the assets, but may be interested in the assets. In venues where only lump sum bids are accepted, they may pass on promising patent portfolios entirely. Furthermore, private sale engagements may receive more attention from the brokers themselves. Multi-lot auctions tend to have several dozen lots in each auction. Online venues may list even more at any given time. Private sale brokers, on the other hand, only handle a few deals at any given time. This extra attention may translate into better, more targeted marketing materials, or more focused contact with prospective buyers regarding the patent portfolio. The principal disadvantages of the private sale process are selectivity and pricing. Because private sale engagements take so much time, a patent broker can only handle a certain number at any given time. This fact makes it harder to commence a private sale engagement with a broker than either listing with an auction house or an on-line venue. A private-sale broker may also charge a retainer in addition to a commission.

When faced with these options, a seller should strongly consider the complexity or the portfolio and the likely hurdles to completing a sale. For large, high-value patent portfolios, a private sale makes the most sense. These are portfolios that require a lot of diligence before the deal is completed, and the private sale platform insures that enough time will be allotted to conduct this. Also, using a more flexible bidding structure may allow the seller to maximize the sale price or find the optimum terms structure for the deal. The extra cost that the seller may incur by going through with a private sale may be well worth it to ensure that the deal has the maximum likelihood of success. In contrast, the auction platform tends to be best suited for small- and medium-sized portfolios. The main issue is time for buyers to conduct their diligence. Assuming that these portfolios do not require onerous diligence by a prospective buyer, then the time allowed for diligence in an auction platform is sufficient. The auction platform, much like private sale, allows for extensive marketing of the patent portfolio by the auctioneer. Finally, the online venue is a good choice for lower value and simpler patent portfolios; however, it may not be suitable for more complicated, valuable patent portfolios. Online venues normally require the least amount of money upfront. Also, depending on the online broker, the marketing of the portfolio may be limited to the listing and basic seller support throughout the process. As of 2010, standard total commissions for auctions and private sales are between 20% and 40% with an online venue charging less (and, sometimes, considerably less).

When the seller decides to put a patent portfolio up for sale, it should consider whether it wants to retain a license to practice the claimed inventions of the patents in the portfolio. This license will be a non-transferable, non-sublicensable, non-exclusive license to the seller of the patents. It will be included as part of the patent sale agreement between the buyer and the seller. This arrangement (often referred to a "license-back") is strongly recommended for any seller that is an operating company.

The chances of success in a patent sale are greatly improved when the seller is aware of the issues surrounding sale of its portfolio and can match the requirements of its portfolio with the appropriate platform. It is also in the best interests of the seller to find the right match on its first try. Once a portfolio has been taken to market and not sold, it can be difficult for any broker using any platform to reignite interest in the portfolio with prospective buyers.

BUYER DILIGENCE AND CLOSING

There are two main tracks to buyer diligence. The first track is evaluating the patent portfolio on its technological and legal merits. This type of diligence involves the prospective buyers analyzing the strength of the claims in the portfolio and the market which they address. This may involve, for example, a detailed examination of the patents' prosecution histories and an evaluation of what product(s) are on the market that currently practice one or more of the patents' claims. Furthermore, the buyer examines how the portfolio for sale fits in with its larger holdings and whether it makes sense for the buyer to acquire the portfolio and if so, at what price.

The second track of diligence is where the buyer examines issues surrounding the title of the portfolio. When performing this second type of diligence, the buyer makes sure that there no outstanding issues surrounding the portfolio that would undercut the buyer's future right or claim to ownership of the assets.

In a private sale, the first track is normally undertaken first by the buyers to determine if they have any interest in the portfolio. Once the buyer has established an interest in the portfolio or won the bid, it starts on the second track of diligence. In an auction, because of the time constraints, both diligence tracks are performed more or less concurrently.

The two documents used to transfer patents from one owner to the next are the assignment document and the patent sale agreement. The assignment document is a document filed with the U.S. and Trademark Office that updates ownership of patents and applications with the office. The sale agreement is a contract between the seller and the buyer that establishes the terms of the sale. While the format and terms of the contract differ from deal to deal, they all include these parts: seller identification, buyer identification, assets at stake, and terms of the sale. The contract is executed by both the seller and the buyer. The assets at stake section explicitly names the assets to be transferred in the sale. The list of assets can include U.S. patents, international patents, U.S. and international patent applications, Patent Cooperation Treaty (PCT) applications, and provisional applications. The list may also include applications that were abandoned or expired patents that might be legally related to one or more assets in the portfolio. Typically, there is catch-all language that makes it clear that the buyer will take possession of all assets on the list as well as assets, abandoned or not, U.S. or international, that are legally related to any asset on the asset list. These provisions are important to most buyers because they want absolute control over the entire patent families in the portfolio, not just the individual patents.

The last section in the sale agreement lists the terms of the contracts. This section can vary widely between contracts. Along with stating the sale price for the assets and the closing date of the transaction, this section also includes "Representations and Warranties." The representations and warranties section typically includes clauses that clearly state that the seller is the sole legal owner of the assets, has the right to sell the assets, and that the assets are clear of any liens, mortgages, and are no longer acting as security for any outstanding loans. This section will also have language stating that the assets have no encumbrances, such as previous licenses. If licenses do exist, there is a separate section which will name each license as an exception. Other language in this section generally may state that all documents provided by the seller to the buyer are truthful. Also, language stating that appropriate fees have been paid or that the patents have been properly maintained and that the seller has never received a judgment by a court or by an administrative body such as the U.S. Patent and Trademark Office that any member of the patent portfolio has been found invalid may be included. Other clauses restating or expanding upon these ideas may also be present in the representations and warranties section. The clauses listed here are not exhaustive, and other provisions may also be included as part of the representations and warranties section.

The clauses in this section generally insure that the seller is rightful legal owner of the assets to be sold and that the seller can legally transfer them to the buyer. In many ways, the types of representations and warranties in a patent sale agreement are not that different from those in other sales contracts, such as the sale of businesses or real estate. The seller should familiarize itself with each clause and make sure that it is comfortable with all the clauses.

SELLER DILIGENCE ITEMS

While diligence issues normally arise for the seller at the end of the process, it is important for the seller to have a handle on the expected issues at the beginning of the process. The following steps are a non-exhaustive list that will help all the parties involved to expedite the close of the deal. Once the patent holder has made the decision to sell, it should make a list of all assets and any related assets that it wishes to sell. Related assets are domestic and international patents or patent applications that claim priority to assets in the patent portfolio. Again, this list should be inclusive and include related assets that have been abandoned or expired. The seller should locate the prosecution histories for at least the domestic patents (and related applications) so that they can be sent to prospective buyers, if necessary. If any U.S. patents have been through reissue or reexamination, this fact should be noted and the case histories should be located. For foreign applications, if the patents have undergone opposition, this fact should be noted and case histories located for further reference. The patent holder should make a list of all licensing agreements and locate the contracts. If any of these agreements include confidentiality agreements, it should contact the licensees and seek limited release to discuss these agreements with prospective buyers. At this point, it is wise to make sure that no terms in these agreements would prevent transfer of ownership. If the seller itself is not the original owner of the assets (the original inventor or original assignee from the inventors named on the patent), it should locate all prior transfer agreements and assignment documents. The patent holder should make sure that all prior liens have been removed from the assets. Furthermore, the owner should make sure that all information available from the USPTO is correct. Discrepancies between the owner's records and the patent office can cause issues as the prospective buyers undergo their due diligence. Specific issues of which sellers should be aware pertain to inventorship, priority, and transfer of assets. The patent holders should make sure that inventorship for all patents and applications is correct and that each inventor signed assignment documents assigning away all interest in the patent or patents. The patent holder should check and make sure that all claims of priority have been made and that they were registered with the USPTO. Any problems should be corrected with the USPTO as soon as possible. Furthermore, any and all maintenance fees should be paid up and kept current throughout the entire process until the effective date of the patent sale agreement.

Transfer of patents in the U.S. is essentially a two track process. One track is the registration of the ownership performed with assignment documents filed with the USPTO. The other track includes sale and transfer agreements governed by the local laws of the entities involved. The patent holder should ensure that at each prior transfer leading up to the present deal that documents were properly assigned from one entity to the next, and that any corresponding legal documents required by applicable jurisdictions were executed and filed properly.

CONCLUSION

The creation of a patent market has been an interesting development for the field of intellectual property. This market has helped patent holders monetize their intellectual property without resorting to the legal system. Furthermore, the patent market has become an important way for buyers to develop their portfolios quickly and supplement their own patent application filings. The market for patents, though, is still in its infancy and poses significant challenges to participants. In my experience as a broker for an intellectual property auction house and a private sales consultancy, the best outcomes come to those sellers who take a hard look at their patent holdings and are realistic about the potential outcome of a sale. While a patent sale is relatively straightforward compared to other avenues to revenue, properly vetting the portfolio for salable assets and positioning the assets on the market requires expertise. Successful sellers dedicate their own resources to the project (by either using in-house professionals or working with outside consultants) and work with patent brokers to maximize their outcomes in the marketplace. The patent market is now firmly established and the sales process should always be considered for patent holders looking to generate revenue from their patent holdings.

CHAPTER 12

Patently Ridiculous

Gerry Langeler

Gerry Langeler joined OVP Venture Partners in 1992 and focuses on investments in the digital energy, software, imaging and digital biology sectors. His investments include Captivate Network, Carbonflow, Elekom, EnerG2, Foundstone, Max-Viz, Now Software, Portland Software, Preview Systems, Surplus Software, TView, Unify, Vascular Solutions, Viral Logic Systems Technology (VLST), and Webridge. He currently serves on the Board of Directors of Carbonflow, Collaborative Software Initiative, CradlePoint, EnerG2 and Max-Viz and is a Board Observer to DataSphere, Serus, Tigo Energy and Viral Logic Systems Technology (VLST).

Prior to joining OVP, Mr. Langeler was co-founder of Mentor Graphics Corporation (NASDAQ: MENT) in Portland, OR in 1981. There he served as President, and over eleven years helped lead Mentor through its IPO to over $400M in worldwide sales and over $1B in market capitalization. He began his technology career at Tektronix.

Mr. Langeler received his AB in Chemistry from Cornell University and his MBA from Harvard University. He is the author of *Take the Money and Run! An Insider's Guide to Venture Capital* (LuLu Enterprises, 2011), as well as *The Vision Trap* (Harvard Business Review, 3/92). Mr. Langeler serves on the Advisory Board of Highway 12 Ventures in Boise, Idaho. Mr. Langeler can be contacted at langeler@ovp.com.

For startups and small businesses, patents are unimportant. Except when they are not! Is that clear?

Seriously, what we have found over many years in the venture capital business, backing literally over one hundred companies, is that the value of patents varies widely. In this chapter, we will try to outline some of the parameters that weigh on their value, to help you decide how much emphasis to put on them in your company.

However, first a disclaimer: the author of this chapter is not a lawyer. For good legal advice on this subject, find yourself a good intellectual property lawyer.

So, how can we say that patents are ever unimportant? Because we have seen company after company spend many thousands of dollars to document, file and sometimes be awarded patents that neither deterred competitors, nor attracted acquirers. Perhaps this reflects the bulk of our portfolio over time being in the information technology space. Much of the value-added in that arena has been in software, and software has proven to be a very murky area as far as the value of patents. Since software is by its very nature highly malleable, it is remarkably easy to find other ways to accomplish the same result, and so working around patents is reduced to an art form by clever engineering teams.

So, how can we say that patents are sometimes very important? Because we have seen other companies either live by their patents, or die by other people's patents. The overwhelming majority of those companies have been either in the medical arena, or more recently in clean technology. And their patents have tended to center on composition of matter (such as new drugs), or techniques to efficiently create certain compositions of matter (such as catalysts or enzymes).

Perhaps the most efficient way to cover this subject is to share with you a number of real-world examples (occasionally disguised to protect the innocent) and then draw some more generalized conclusions from what we observed.

INVENTOR, STEAL THYSELF

Let us start with the wild-haired inventor who knocks on our door looking for venture capital. He has his patent in hand, and based on that assumes we will fall all over ourselves to give him money. Not very likely. It turns out that patents by themselves are rarely valuable. Only when they are reduced to products, and the products are part of a company, and that company can compete in substantial markets, do those patents have value to people like us. A simplistic way to think about this is "ideas are cheap, but execution is precious."

Of course, this is not always true. Sometimes a core patent is worth a lot of money on its own if it stands in the way of major industry giants, and in turn they are willing to pay a royalty for its use. But, companies built on royalty streams are notoriously hard to get to scale, and notoriously hard to get liquid for their investors. They may be very nice income-stream vehicles for friends and family, but not likely to attract significant institutional capital.

So, when do you know which you are? Perhaps the easiest way is to see what "offers" come your way when you go out to raise money with your invention / patent. I know of an entrepreneur who, in my judgment, played this exactly wrong.

He had a patent that may be fundamental to an important industry and was trying to raise money for a startup to commercialize that technology. Because he had no team around him, and no business experience to give credence to his propensity to execute, venture capitalists held back. During that process, he was approached by a large industrial player in the field, offering to buy the rights to the patent.

They offered him $10 million. And he turned them down!

Here is why I think he is crazy. First, $10 million is a life-altering sum for most of us. To say "No" to that means you have to be remarkably confident that the probability of a much larger pay-day is high. But it is not. Let us do the math…

If he is successful in getting his company funded by institutional sources, by the time he has raised all he actually needs (probably at least double what he thinks he needs), and hires the management team he needs, and gives them an equity stake in the firm, he will more than likely own something in the 5%-10% of the enterprise. That may seem small to you, but that is the way the world usually works.

So, for him to be indifferent to the $10 million offer today, he will need to see an enterprise value at some future liquidity event of $100 million - $200 million. Is that possible? Certainly! But, is it highly likely? Not a chance. I can tell you that as a venture investor, we back fewer than 1 out of 100 opportunities we see. Of those we invest in fewer than 1 in 5 end up with a triple digit millions enterprise value. So, we are talking a less than 1 in 500 chance that his "bet" to forego the $10 million makes economic sense. That is entrepreneurial zeal (or blind spot) in the extreme.

WHAT IS OLD IS NEW AGAIN

OK, let us move past the crazy inventor to a more established business. In this case, the company was in a capital equipment sector, with a series of patents around process and materials technology that could yield a step function in productivity for potential customers, if it could be commercialized. The company raised multiple rounds of venture capital to pursue this goal, totaling in the tens of millions of dollars, and had the good fortune to have a principal inventor who was over the age of 65.

Now, if this is news to you, consider the next few sentences worth the price you paid for this publication. A little recognized part of the patent law can offer you tremendous potential dividends. It seems U.S. patent law and Social Security share a common attribute. Both set their age metrics back when 65 was the average life expectancy, and have not really adjusted very much to better health care and better living conditions in modern society. In the past, the patent office did not want to be issuing patents to people who died waiting for them to act. So, if you file for a patent and you are over 65, that application goes right to the head of the line! Instead of a multi-year wait to even get an office action, you might be looking at multiple months to get a patent fully issued. It is a case of "old guys and gals rule."

So, this company with the "old" prolific investor was racking up patents like mad. And some of these patents were very fundamental, broad claims. But at the same time, it was having difficulty getting past the "R" stage of R&D to the "D" stage, much less getting to a finished product that could be sold and supported.

Finally, the venture investors began to run out of patience and willingness to continue to fund the beast. A consultant was hired to estimate how much the patents were worth, in case the best answer was a shut-down with a sale of the intellectual property. The result was a shocker. After tens of millions of dollars invested, and the issuance of many tens of patents, the "salvage" value of the IP came in at slightly over $1 million.

If there ever was testimony to the point that "ideas are cheap but execution is dear" this is it. Had this company been able to take the next step, and really commercialize some of those patents and been able to demonstrate customer demand for the resulting products, that intellectual property might have risen in value by 100-fold.

WHAT DO YOU CALL 1000 LAWYERS AT THE BOTTOM OF THE SEA? A GOOD START.

Apologies to my lawyer friends, but that old joke reminds me of a truism for young and/or small companies. Remember if you are basing your company value on the value of your patent portfolio, said value is only as large as your ability to enforce your position. Put another way, if you plan to go to war (aka court) against an opponent who has more lawyers on staff than you have employees, or has more cash on hand than you can count, be prepared to lose. It does not matter how strong your position is, or how certain you are of ultimately winning. You have to survive the battle to win the war. Most small companies do not.

Instead, think of your IP position as trading stock. If you are being infringed upon, look to do some horse-trading with the offender before you look to slap a lawsuit on them. It may not work, but more often than not, I have seen this play out to the benefit of everyone who is not billing by the hour. Often, to avoid litigation you can get some useful IP from the other guy, and maybe even a one-time cash payment or on-going royalty. And it is often the case that while the other side may be infringing to some degree, even with a license to your patent, they are missing something they need to fully exploit it. Your inventor(s) who know where the bodies are buried along the path to the patent filing may have proprietary knowledge on how to implement that technology that keeps the other side at bay. In fact, years ago I was told that the best way to kill a competitor in the software business is to give them your source code. Without the engineers who wrote it, most companies break their pick trying to figure out how it actually works.

But wait. We have all seen published reports of companies that have extracted significant sums from large competitors who were trampling their protected positions. When do you know if it is time to make war rather than make friends?

One circumstance I have witnessed where it makes sense to act is when cash is sitting there to be had, and people are willing to pay a "small" portion of it to take home the rest of it. A good example is the escrow portion of M&A transactions.

We had a portfolio software company with some proprietary IP that appeared to have been obtained improperly by a competitor. Neither firm was particularly strong financially, but the other firm was definitely more substantial. If we sued them, we might both go down in flames as the lawyers battled it out. But then a miracle occurred. They got acquired by a company with deep pockets, and their venture capital investors were looking at a significant payday. But, as with all such deals, a portion of the transaction was held back in escrow against any potential claims against the assets of our original competitor.

Our company was happy to oblige. It filed suit against the new parent firm, with the request for monetary damages in an amount that fit within the escrow hold back. Since the venture investors on the other side had already received 80% of the deal value, they were more than happy (OK; probably not happy, but accepting) to part with some of that remaining 20% to release the remainder for themselves. In this case, that "small" portion of the escrow amounted to about $10 million, and funded our company for years.

The take-away here is that sometimes *when* you choose to defend your intellectual property can be more important than *whether* you choose to do so.

NOW YOU SEE IT, NOW YOU DON'T

After a few decades in the venture business and another on the start-up side of the street, I thought I had seen everything when it came to patent issues. Not so.

We had a company in the biotech arena that lost a key piece of value of a patent because it did not proofread the patent application carefully. This was a company with a terrific IP law firm, an in-house counsel steeped in the art, and a set of world-renowned scientists. So, all of them knew the drill and it was certainly not the first time they had dealt with the complexities of filing patent applications.

Everything was going swimmingly until the company was in

serious negotiations with a major pharmaceutical company about licensing this IP. Of course, the big firm put their team on the diligence process, with fresh eyes.

If you follow this field even casually, you know that all DNA consists of the four nucleotide bases—adenine, cytosine, guanine and thymine—usually represented in the literature by the their respective first letters: A,C,G,T. And if you know that (or do now) you may also know that the strings of these (in differing order) to describe a particular biological compound may extend into the many, many thousands. AGTCTTGTCAATGGACCAGTTAACCGTAATAA...

Sadly, during the drafting of the patent application, someone transposed two of those letters, and it was not caught by anyone looking over the documents. The patent office actually has a period of time after you file during which errors can be corrected, but that clock had run out. So, when the potential strategic partner noticed the error, it was too late to fix it, and the intellectual property of the compound was essentially now in the public domain.

So, even when you are sure you have a patent, you may not! More commonly, it is amazing how often we find that patents our firms have can be invalidated (or at least called into question) by prior art elsewhere. We do that to competitors as well, of course. Just know that for all your good efforts and sincere research to prove you have a unique idea, there are bright people out there who will dig and dig to prove you wrong. Sometimes, they are even right.

HERE COMES THE JUDGE

I mentioned at the beginning of the chapter that we have seen companies live and die by patents. Here is a case in point.

Since this is all on the public record now, and is far enough in the past to tell, we will name names. The company in question was CellPro, a biotech equipment firm in Seattle with technology coming out of the Fred Hutchinson Cancer Research Center there. The firm had developed technology to aid in bone marrow transplantation. Historically, those about to undergo such a procedure for cancer treatment had a bunch of marrow extracted from their hip and then were blasted with radiation or chemotherapy or both to kill the cancer cells in their body. But that also killed off all the good cells in the remaining bone marrow. So, after the harsh treatment, the extracted bone marrow was reintroduced into the patient, where after a scary period where the patient had essentially no defense to infection, the stem cells in it regenerated and hopefully led to a full recovery.

The only problem was, in extracting that bone marrow, cancer cells came along for the ride. So, the doctors always were re-engrafting not just healthy cells but unhealthy ones. CellPro had a column that the extracted marrow went through that separated the good cells from the bad, so only good cells were re-introduced into patients. This was clearly a life saving, life changing technology.

CellPro had a successful public stock offering on NASDAQ and was steaming along when hit by a patent infringement lawsuit by a very large firm. Of course, the suing company set the jurisdiction for the case in their own backyard. To make a long story short, even in the face of what CellPro thought was compelling evidence that their IP was valid, the judge ruled against them. The supposedly infringed big firm made our company stop selling the product, even though that firm had no competitive offering in the market. People died as a result of the ruling. So did CellPro.

Now whether the judge was right, or honestly wrong, or something else was going on behind the scenes, we will never know. But it was a cautionary tale for sure. The only good news, well covered in the *Wall Street Journal* at the time, was that ironically just before the company had to cease operations, its CEO was diagnosed with cancer, and needed a bone marrow transplant. He used his own product to save his life—and it worked!

The lesson we learned here, and I hope you will find useful, is recognize that judges are people. They can be as right or wrong as any of us and perhaps even "influenced" in ways we would rather not think about. If at all possible, find a way to settle whatever dispute you have over patents before you submit to that all or nothing outcome.

WALK SOFTLY, CARRY A BIG STACK OF PAPER.

OK, enough with the horror stories. Here is one that demonstrates just how valuable good patents can be. We are involved in a company that has a terrific patent portfolio, and one patent in particular sits squarely on the head of its arch enemy. The enemy, to be fair, has some interesting patents of their own and while we are quite sure we should win any battle, we still remember the CellPro lesson and would like to find a better way to exercise our options. And the arch enemy also has some very interesting products (some of which infringe) farther down the development curve than some of ours.

So, the CEO of our company has gone out and rounded up an exclusive license from a third party to yet another piece of intellectual property that will make life exceedingly difficult for this competitor. Now it is time to apply the squeeze. The competitor is about to have a major transitional event, for which it needs a clean IP situation. But they do not know about our firm's new license, nor do they seem to understand how the patent we had issued already has them in a defensive mode (or if they do, they certainly do not want to signal that).

We have another "transitional event" in mind for them. Perhaps we "make them an offer they cannot refuse." We simply give them a choice: either we merge the firms, with our company being the surviving entity, or we file suit and make their lives a living hell! As they also said in *The Godfather*, "It's nothing personal, just business."

THE CIRCLE OF LIFE

There you have it. You have gotten at least a quick pass at how patents can be useless, useful or all powerful. You have seen how and when they are used can be as important as if they are used. And you can see how, depending on the circumstances, they have very different value in the eye of the beholder.

My closing advice is to think hard about what you choose to patent, and think about patents as much as defensive weapons as offensive ones. And if you have inventors over 65 on your team, treat them well!

About the Editor

David B. Orange practices in the Electrical Patent Group at Buchanan Ingersoll & Rooney. He has led M&A due diligence, litigation and product clearance teams. He has managed patent portfolios both as outside and as in-house patent counsel and has secured key patents for clients. David has performed research on hardware and software for a variety of industrial and academic labs and his semiconductor designs have won a number of grants and honoraria. He earned a B.S.E.E., J.D., and LL.M. from the University of Washington.

David serves on the Boards of Directors for the University of Washington School of Law Alumni Association and the MIT Enterprise Forum of Washington, DC, where he leads the "Talk to the Angels" series. He is as a commercialization grant reviewer for the National Science Foundation's Small Business Innovation Research Awards.

Related Titles from Logos Press®

http://www.logos-press.com

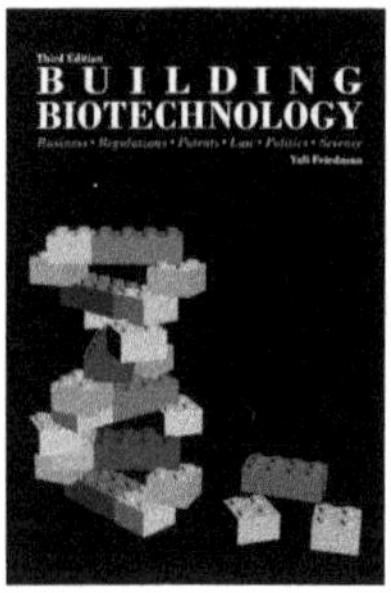

Building Biotechnology

Scientists know science; businesspeople know business. This book explains both.

Hardcover ISBN: 978-09734676-5-9
Softcover ISBN: 978-09734676-6-6

Best Practices in Biotechnology Business Development

Valuation, Licensing, Cash Flow, Pharmacoeconomics, Market Selection, Communication, and Intellectual Property

ISBN: 978-09734676-0-4

Biotechnology Entrepreneurship

From Science to Solutions

Hardcover ISBN: 978-1-934899-13-7
Softcover ISBN: 978-1-934899-14-4